DIY Drones for the
Evil Genius

Evil Genius Series

DIY Drones for the Evil Genius
Design, Build, and Customize Your Own Drones

Ian Cinnamon

Romi S. Kadri & Fitz Tepper

New York   Chicago   San Francisco   Athens   London   Madrid
Mexico City   Milan   New Delhi   Singapore   Sydney   Toronto

Cataloging-in-Publication Data is on file with the Library of Congress.

McGraw-Hill Education books are available at special quantity discounts to use as premiums and sales promotions, or for use in corporate training programs. To contact a representative please visit the Contact Us page at www.mhprofessional.com.

DIY Drones for the Evil Genius: Design, Build, and Customize Your Own Drones

5 6 7 8 9 10 QVS 22 21 20 19 18

ISBN 978-1-25-986146-8
MHID   1-2-5986146-5

The pages within this book were printed on acid-free paper.

Sponsoring Editor
Michael McCabe

Editorial Supervisor
Donna M. Martone

Project Manager
Poonam Bisht, MPS Limited

Acquisitions Coordinator
Lauren Rogers

Copy Editor
Kirti Sharma Kaistha, MPS Limited

Proofreader
Rashi Sinha, MPS Limited

Indexer
Edwin Durbin

Production Supervisor
Lynn M. Messina

Composition
MPS Limited

Art Director, Cover
Jeff Weeks

Contents

About the Authors

Ian Cinnamon is an engineer and entrepreneur determined to bring ideas to life. At a young age, he taught himself computer science which led to him authoring the Scientific American Book Club bestseller *Programming Video Games for the Evil Genius*. He has been a leader and advisor at a number of Silicon Valley-based technology companies. In addition to serving in product roles at Apple, Yelp, and Zynga, Ian led his startup, *superlabs*, through acquisition and serves as Director of Strategy at the nonprofit Immunity Project. A Forbes 30 Under 30 Honoree, Ian is an MIT graduate and MBA candidate at Stanford Graduate School of Business.

Romi S. Kadri began his engineering career at Rolls-Royce, manufacturing compressor blades for jet engines before heading to MIT where he graduated in Engineering and Entrepreneurship. Having founded and invested in numerous technology businesses and served in everything from Artist Management at Interscope Records to Innovation at Sonos, he continues to fulfill his love of bringing the most impactful technologies and music to the world. In 2016 Romi became a certified pilot, turning his childhood dream of flying airplanes to reality.

Fitz Tepper writes for the technology news site TechCrunch about a wide range of topics ranging from FinTech and Bitcoin to drones and self-driving cars. Fitz also currently attends Fordham University's School of Law, and is a graduate of USC's Marshall School of Business. Fitz is also a frequent technology commentator on CNN's Los Angeles-based "CNN Newsroom."

Acknowledgments

THANK YOU TO MY PARENTS Karen and Ken for giving me my first drone and encouraging me to fly higher and higher (not just with drones, but with life!). Thanks to my ever-supportive sister Molly for liking all of my drone photos on Instagram and encouraging me to turn my obsession into a real book. Thank you Ida and Bach for putting up with the ever-expanding drone collection (we seriously need to build a drone hanger). And Romi, Fitz, Katie, and Shane—without you guys, this book would still be a crazy idea floating around in our heads. Thank you for helping make it a reality!

Ian Cinnamon

I send a million thanks to Ian Cinnamon for presenting me with this wonderful opportunity to share with you my immense passion for aviation and the physics of flight. To my parents, siblings, grandparents, Aden and the Santa Barbara Kadris, Kevin Rustagi, Christian Bailey, Billy Thalheimer, Matt Berlo, Ian Swan, Prof. Doug Hart, Prof. Peko Hosoi, and Dr. Barbara Hughey: collectively, you have inspired me and trained me in everything I needed to share my passion for flight with everyone who picks up this book.

Romi S. Kadri

Thank you to all of the early computing pioneers who didn't listen to the people that said technology was just a temporary fad. The world is a better place because of you.

Fitz Tepper

Foreword

THE WEEKEND BEFORE I READ *DIY Drones for the Evil Genius,* I was off-roading in my Jeep toward a favorite fly-fishing spot in the Rocky Mountains. We came to a stream. Before the trip I had designed and 3D-printed a special mounting device, which now anchored my favorite quadcopter drone to the Jeep via an electronically controlled magnet. As we approached the stream I released the magnet and launched the drone. The payload on this adventure was a high-resolution video camera, and I sent the drone aloft to record what would either be a successful traverse of a rushing mountain stream, or a spectacular aerial video recording of my final moments on earth.

We made it. Later, I swapped out the camera from the drone's platform, replaced the payload with a significant-size water balloon, and dropped it with satisfying effect on an unsuspecting family member.

Does this qualify me as an Evil Genius? Perhaps. But safely back home, having read this book, all I could think was: Geez, I wish I had read this before I started messing around with drones; it would have saved me at least a couple of years of trial-and-error experimentation.

You don't have to be a genius (evil or otherwise) to benefit from this book, but it definitely helps to have an enthusiast's spirit and a desire for hands-on learning about one of the major disruptive technologies in 21st century. Drones big and small are already being used to conduct reconnaissance and deliver medical supplies to disaster areas, deliver parts inside warehouses or on campuses, monitor crops and climate change, and even to deliver Internet access to remote areas of the planet.

Drones are not all work and no play, of course. Racing, aerial acrobatics, exploring, photography, and videography are just a few of the recreational uses of these versatile flying robotic platforms.

Yet at HP Labs, where we create disruptive technologies like 3D printing and immersive computing, we recognize that we're very much at the beginning of the drone era. It's like the early days of the PC industry. There's unlimited opportunity for anyone to design, create, build, and customize drones for almost any application they imagine. The basic component parts are easily available; maybe someday soon they'll even be delivered to your door by drones. The few essential ingredients that might be lacking are knowledge of the basics of flight, some guidance on how the parts work and how to put them together, and some tantalizing ideas what you might want to do with your own personal drone.

Those missing essential ingredients are supplied in this book. Ian Cinnamon and his coauthors Romi S. Kadri and Fitz Tepper have crafted a comprehensive primer for the drone enthusiast and for those who think they might want to become one. *DIY Drones for the Evil Genius* starts with the basics of flight and aeronautical engineering, which might be a deep dive if you're merely interested in buying an inexpensive toy drone to fly around the house or office to mess with household pets and coworkers (in fact, the authors recommend starting with a cheap, palm-size toy drone to master flight and control fundamentals). For those not quite ready or lacking the time for the Do It Yourself adventure, there are descriptions and product recommendations for the most popular store-bought drones on the market. And there are lots of them!

But for the intended audience of do-it-yourselfers and makers—for innovators who wonder, *"Wouldn't it be cool if I could do this? Or maybe that?"*— *DIY Drones for the Evil Genius* captures the spirit of invention and reinvention that is propelling the whole drone revolution, and backs it with proven evil genius know-how.

Shane Wall

Chief Technology Officer and Global Head of HP Labs

Palo Alto, California, USA

As CTO of HP Inc. and Global Head of HP Labs, Shane Wall drives the company's technology vision and strategy, new business incubation, and the overall technical and innovation community.

Introduction

WHAT TECHNOLOGIES HAVE FUNDAMENTALLY CHANGED the way you live? Take a look at the Internet and cell phones: they fundamentally altered how you go about your daily routine. And while they weren't widespread even 20 years ago, today it is practically impossible to live without your iPhone or Google. But what's next? What technology is being developed today that will soon become an essential part of our daily lives? While there are many candidates, you'd be hard pressed to find a technologist that doesn't believe that drones, or unmanned aerial vehicles, will soon fundamentally change daily life as you know it.

Sound overly dramatic? Perhaps you just think of a drone as the expensive holiday toy you use to fly over houses and take videos and pictures of the neighborhood. But drones are so much more than that. In this book, you explore exactly how drones are poised to be the technology that will most impact the world over the next decade. You'll learn how drones will revolutionize dozens of industries including delivery, construction, and even personal transportation.

Don't worry—this book is not a history lesson. The best way to learn is by doing, so this book is a hands-on guide to learn how to build your own drones from scratch. You'll also learn the basics of flight, an introduction to aerospace engineering, how drones work, and everything there is to know about the current state of the drone industry.

DIY Drones for the Evil Genius starts with a deep dive into aerospace engineering, which is the science behind what makes drones actually fly. You'll also explore all the different parts that go into building your own drone. Next, you'll learn the different types of drones (there are a ton!) with a focus on four-propeller quadcopters. Before actually building, you'll learn about the Federal Aviation Administration, and what you need to do to legally and safely fly in the United States.

The second part of the book teaches you how to fly your own drone, using a small "micro-drone" purchased online. After mastering that, you will move into the third part of the book, which explains what parts you need to build your own drone, and what each component does. Then, in section four, you start building! This section will walk you through every step of building your own personal drone, making sure to explain everything along the way.

After building your own drone, the book teaches you how to do some high-performance (and more complicated) tricks with your quadcopter. Learn how to attach things like lights and cameras to your drone, as well as how to film professional looking footage from your craft. You'll even learn how to broadcast live from the air to viewers around the world, essentially letting you create an amateur news helicopter!

By the time you're done reading this book, you will be an expert on all things about drones. So strap in and get ready to learn everything there is to know about drones, the drone industry, and why these machines will soon change the world.

DIY Drones for the Evil Genius

SECTION ONE

Preflight Checklist

Time to head to the classroom. This tutorial section provides a systematic overview of aerospace engineering with a specific focus on drones. You'll learn about aircraft parts, control mechanics, and best safety practices.

PROJECT 1

Drone? Aircraft? Unmanned?

Aerospace engineering 101: Just like a classroom but way more fun. Learn the different types of flight, with a focus on airplanes, helicopters, and quadcopters.

Since you're reading this book, you're probably excited to build your first drone … and you'll get to that soon! But before you dive into the intricacies of drones, you should understand the different forms of aviation.

Humans have always had a desire to fly. An obsession that began with airplanes has evolved into spacecraft, helicopters, unmanned multirotor aircraft, jetpacks, and so much more. Aerospace engineering breaks the above list into two separate categories: aeronautical engineering (aircraft) and astronautical engineering (spacecraft). In this book, you'll focus on learning the basics of aeronautical engineering (Figure 1-1).

Even if you know nothing about drones, you've definitely heard of them. In reality, a drone is defined as a robotic vehicle, usually unmanned. Within the context of this book and popular culture, "drone" refers to a multirotor aircraft. In this book, the terms "drone," "multirotor craft," and "quadcopter" will be used interchangeably. These vehicles are almost unmanned and are traditionally comprised of four, six, or eight arms. Each arm contains a

high-performance motor and propeller, allowing the craft to lift significant weight and maneuver around tight spaces (Figure 1-2).

Before you dive further into drones, you should also understand the basics of flight. Over the next several pages, you'll learn how airplanes, helicopters, and drones fly. Throughout the rest of the book, you'll utilize this knowledge and build your own drone to get a deep understanding of how it flies.

> "My flight instructor once asked me, 'what makes airplanes fly?' As an engineer, I started throwing out terms like 'Lift & Drag', 'Gas', 'Gravity', 'Thrust', 'Wings!' After shaking his head repeatedly, he turned to me … *'Money!'*, he said. Sure, manned aircraft are expensive to fly. Thankfully drones are a much more affordable way to keep your head in the clouds."
>
> –Pilot Romi Kadri

The history of aircraft design stretches to the late 15th century when Leonardo da Vinci produced some of the very first human-powered airplane and helicopter designs. Unfortunately,

FIGURE 1-1 Airplane and helicopter.

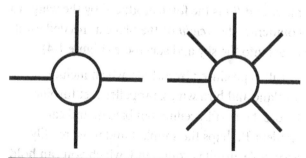

FIGURE 1-2 Different arm configurations.

FIGURE 1-3 Wright Brothers' airplane.

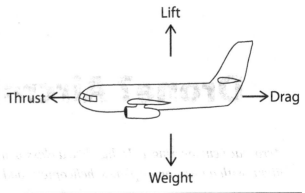

FIGURE 1-4 Diagram of principal forces on airplanes.

his attempts failed as the humans he hoped would power his machines into the sky didn't have enough power to get off the ground. The Wright Brothers' "Wright Flyer" achieved the first successful flight in 1903 in a powered airplane, a biplane with a 12-horsepower gasoline engine, bike parts for gears, two propellers, and a wing-warping control mechanism that matches the same physical principles as the hinged control surfaces found on airplanes today (Figure 1-3).

Airplanes fly as a result of four physical forces: Thrust, Drag, Weight, and Lift. *Thrust* is produced by the airplane's propeller, jet, or rocket, and it either pulls or pushes the plane forwards. Once the plane is in motion, the surrounding air pushes back on it, causing *drag*. *Weight* is simply the force of gravity pulling down on the airplane's mass, and *lift* is the force produced by the wings to counteract the *weight* of the plane as needed for it to get into the sky and stay there (Figure 1-4).

Lift is produced by airfoils, which include airplane and bird wings, propellers, jet turbine blades, boat sails, ceiling fan blades, and car spoilers. Perhaps the simplest and most readily accessible airfoil is your hand, which you can hold in a spade shape and stick out of a fast-moving car's window.

Try this project:

In a moving vehicle, hold your hand flat and stick it just outside the window into the passing air. Start with your palm facing downwards and slowly rotate your hand toward the back of the car. You'll feel the force of lift on your hand as you angle it into the passing air (Figure 1-5). Airfoils are shapes that produce an aerodynamic force, meaning: *aero* (to do with a gas, like the air), and *dynamic* (to do with motion). In other words, wings are a type of airfoil; therefore, they produce the force of lift (and some extra drag) when they move through the air.

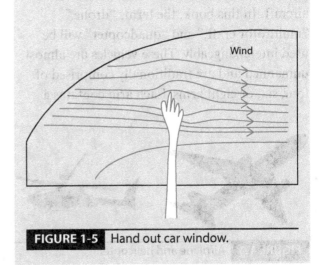

FIGURE 1-5 Hand out car window.

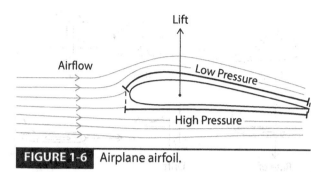

FIGURE 1-6 Airplane airfoil.

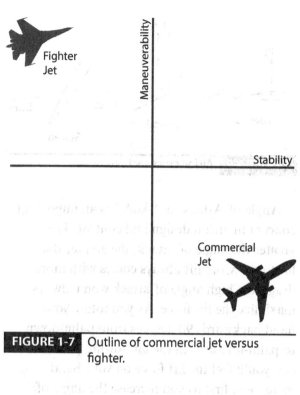

FIGURE 1-7 Outline of commercial jet versus fighter.

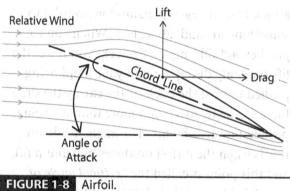

FIGURE 1-8 Airfoil.

Wings produce this lift by creating an air pressure difference between the top and bottom of the airfoil. Whenever a surface has higher pressure on one side than the other, a force exists through the surface in a direction from the high-pressure side toward the low-pressure side. This force is governed by what is known as *Bernoulli's Principle*. An airfoil splits the air into two streams: one over the top of the airfoil, and one under the bottom. The top of the airfoil usually has a curved shape, causing the air to move up and around the airfoil as it passes by. The bottom of the airfoil is flatter than the top, allowing the air to pass directly underneath. Since the air over the top of the airfoil has further to travel in order to pass by the airfoil than the air underneath, it moves faster and is more spread out across the upper airfoil surface. When air spreads out in this way, its pressure drops. With a lower air pressure on top of the airfoil than the higher air pressure underneath, the airfoil produces the upward force of *lift* (Figure 1-6).

And That's All You Need, Right?

No! You Need Controls!

There are two counteracting parts of control to think about: maneuverability and stability. Generally speaking, better stability causes poor maneuverability (think large commercial airplane), while better maneuverability comes with less stability (think jet fighter) (Figure 1-7).

These two control phenomena are the key elements of aircraft design, where the goal of the aircraft designer is to achieve the correct balance of maneuverability and stability for the aircraft's intended use.

On airplanes, aerodynamic control is achieved by changing the shape of their various airfoils. You can feel the effects of changing airfoil shape on your hand sticking out of the car window. The angle between the flat of your hand (which is, roughly speaking, the "*chord*" of the airfoil) and the direction of the wind flowing past the car window (the "*relative wind*") is called the "*angle of attack*" as shown in Figure 1-8.

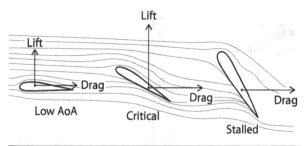

FIGURE 1-9 Airfoil cross sections.

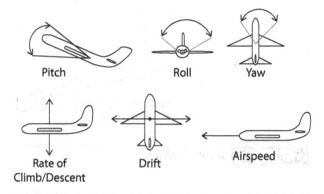

FIGURE 1-10 Aircraft axes of motion.

Angle of Attack, or "AoA," is an important concept in airfoil design and control. The greater the angle of attack, the greater the lift force. More lift always comes with more drag, so a high angle of attack won't always maximize the lift force. As you rotate your hand backwards 90 degrees from palm-down to palm-forwards in the air flowing past your car, you'll feel the lift force on your hand increase at first as you increase the angle of attack from 0 degrees (palm-downwards) to something around 20 degrees. When you rotate just beyond this angle of attack, you'll feel the lift suddenly disappear and your hand being pushed toward the back of the car by the drag force of the relative wind more than it's being pushed upwards by lift. The angle just below this is when the airfoil produces maximum lift, and this point is called the "*critical angle of attack.*" At an angle below the critical angle of attack the airfoil is producing lift. Above it, the airfoil is "*stalled,*" meaning that lift is no longer produced (Figure 1-9).

You might ask yourself, "What does rotating my hand outside the car window like a crazy person have to do with aerodynamic control?" The answer, Evil Genius, is … everything! Aircraft have six axes of motion: *pitch, roll, yaw, airspeed, rate of climb or descent, and drift,* as shown in Figure 1-10.

In all types of aircraft—airplanes, helicopters, and drones alike—each axis of motion is aerodynamically controlled by changing the lift force on the aircraft's airfoils, which can be done

by increasing or decreasing the airspeed over the airfoils. You could also change the airfoils' individual angles of attack using moving parts called "*control surfaces.*"

A control surface is part of an airfoil that can be moved by the pilot or control system to change the shape of the airfoil and thus the angle of attack. Control surfaces can be found on just about every airfoil on an aircraft. Keep in mind that the motion of a single control surface directly affects only the angle of attack of the airfoil to which the control surface is attached. An airplane's airfoils include the main wings, the horizontal stabilizer, the vertical stabilizer, and the propeller or jet blades. A helicopter's airfoils are the blades of the main rotor and the tail rotor, and some helicopters also have horizontal and vertical stabilizers similar to those found on airplanes. Drones have only their propellers as airfoils. Now, you'll learn about the purpose of each of these airfoils, their control surfaces, and the aerodynamic effects of moving them (Figure 1-11).

The primary purpose of an airplane's *main wings* is to produce the biggest chunk of the lift force needed to keep the airplane off the ground. They are the largest airfoils on the airplane, which makes them the perfect platform for a number of control surfaces.

Ailerons are found on almost all modern airplanes to control the axis of roll. They are

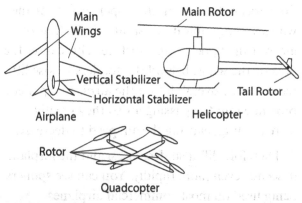

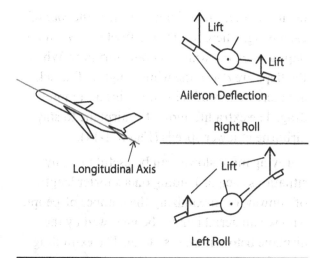

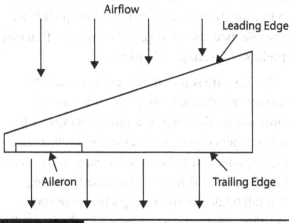

FIGURE 1-11 Airplane, helicopter, and drone airfoils.

FIGURE 1-13 Side-by-side effect of aileron.

FIGURE 1-12 Aileron on wing, showing leading and trailing edge.

typically located near the tips of the main wings on the *trailing edge*—the back of the wing—and move in an opposite direction on one wing to that of the other (Figure 1-12).

To roll the aircraft to the left: the left-wing aileron moves up as the right-wing aileron moves down. This movement causes the left-wing's angle of attack to decrease, reducing the lift force and letting the left wing fall; while the right-wing's angle of attack is increased by its lowered aileron, causing greater lift, and the right wing rises. As the left wing falls, and the right wing rises, the airplane rolls about its *longitudinal axis*—the axis that passes through the middle of the *fuselage*, or body, of the airplane (Figure 1-13).

Additional Information:

Alternatives to ailerons, such as the "wing-warping" mechanism used in the Wright Flyer, do exist. The aerodynamic effect on each wing is identical to that of ailerons, except *warping*, or twisting, the wing in a direction opposite that of its normal position alters its angle of attack. Wing warping is like rotating your hand backwards and forwards to increase and decrease the angle of attack in the air flowing past your car. Imagine if you could reach both hands out of windows on opposite sides and rotate your left hand forwards while rotating your right hand backwards. What would happen? Your left hand would fall with the lower angle of attack and thus lower lift, and your right hand would rise with the greater angle of attack and greater lift (but be careful not to rotate so far as to stall it!). This phenomenon is exactly the same thing as wing warping, but you're rotating your wrists to warp your handy airfoils instead of twisting the beams of the Wright Flyer.

Flaps change the relationship between lift force and airspeed. For a specific shape of airfoil, its lift force increases with airspeed. Flaps change the shape and angle of attack of an airplane's main wings equally on both sides, increasing or decreasing the total lift force on the airplane for a given airspeed. Typically, they are positioned on the trailing edge near the *root* of each wing, where it joins the fuselage. When the flaps are *up*, the wings operate normally. When the flaps are *down*, the wings' angles of attack are increased, producing more lift and more drag. The extra lift allows the airplane to stay airborne at lower speeds (Figure 1-14).

Flying more slowly can be useful in many situations such as landing on a shorter length of runway, or maximizing the number of people who see an aerial banner being towed by the airplane before it moves along. The extra drag is used for deceleration and/or to increase the airplane's *rate of descent*—the speed at which the airplane loses altitude. Usually the flaps can be set to one of several "notches" at different angles between fully up and fully down.

A small amount of flaps are commonly used to take off from a short runway. The small increase in angle of attack causes the wings to generate more lift for the airplane's airspeed, while the flaps are not low enough to increase the drag an enormous amount (unlike having the flaps fully down, where there is a significant amount of drag). The change allows the aircraft to accelerate just as quickly along the runway and become airborne more quickly with the extra lift versus using no flaps at all. Just what you need for short runways!

Spoilers are typically found on larger aircraft and are used primarily to reduce lift; however they are also used to reduce airspeed and sometimes to control the roll of the airplane.

Spoilers are located on the upper surface of the wings and flip upwards, "spoiling" the flow of air over the upper wings' surface. This causes the area of the wings behind the spoilers to lose lift, reducing the overall lift on the airplane. Spoilers protrude into the passing air so they greatly increase drag, causing the airspeed to decrease.

Therefore, lift also decreases, and the airplane descends even more rapidly. You can see spoilers being used on most commercial airplanes immediately after touchdown, "dumping" the wings' lift to keep the plane on the ground and force the tires to gain traction with the runway so that the brakes can be applied as quickly as possible. Because of this, airline pilots often call spoilers "lift dumpers" instead.

Spoilers are sometimes used, particularly in larger airplanes, to assist the ailerons in controlling roll. Imagine a pair of wings with only spoilers (instead of having ailerons at all): if the spoiler is lifted on the left wing only then the left wing's lift force is decreased, causing it to fall while the right wing's lift force stays constant. Such an arrangement results in a roll to the left. In reality, ailerons are almost always used for roll control, and the spoilers merely help them along.

A *horizontal stabilizer* looks like a smaller pair of wings, most often found at the tail end of an airplane (but sometimes at the nose), and allows for *pitch* control (Figure 1-15). Pitch is the rotation of the airplane about its lateral axis—the axis that passes through the side of the

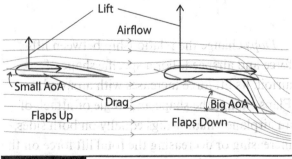

FIGURE 1-14 Flaps.

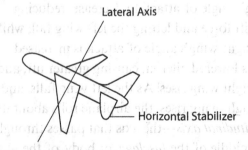

FIGURE 1-15 Horizontal stabilizer.

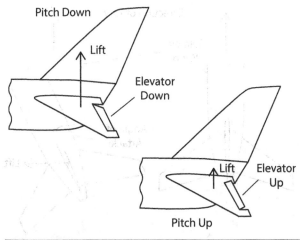

FIGURE 1-16 Elevator up/down.

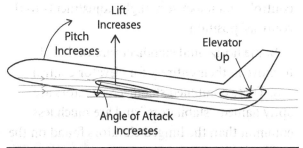

FIGURE 1-17 Angle of Attack (AoA).

airplane at right angles to the fuselage. Pilots are often heard saying "pitch up" or "pitch down" meaning to lift or drop the nose, respectively.

The *elevator* is the main control surface found on a horizontal stabilizer and is typically located on the stabilizer's trailing edge. Much like a single aileron, if the elevator is raised, the angle of attack decreases, causing the horizontal stabilizer to lose lift and fall. If the horizontal stabilizer is at the tail of the plane this causes the nose to rise (*pitch up*). If the elevator is lowered, the angle of attack increases, the lift increases, and the stabilizer rises, causing the nose to drop (*pitch down*) as shown in Figure 1-16.

A horizontal stabilizer can be located at the airplane's nose, as found on the Wright Flyer among other more modern aircrafts. This type of stabilizer is called a *canard*, and its elevator motion is opposite to that of a tail-mounted horizontal stabilizer. Raising the elevator causes a lower angle of attack, reducing lift, and dropping the nose. Lowering the elevator causes a higher angle of attack, produces more lift, makes the nose rise, and pitches the airplane upwards.

So what causes an airplane to climb or descend? You guessed it ... pitch! However, it's not quite that simple. At the moment an airplane pitches up from level flight, it's still moving forwards and hasn't yet started climbing. Increasing the pitch causes the *main* wings to have a higher angle of attack, and you now know how that goes: higher angle of attack means more lift, and *voila*: the airplane begins to climb. Conversely, pitching downwards lowers the main wings' angle of attack, reducing lift and causing the airplane to descend (Figure 1-17).

"How come planes are pitched up when they come in to land?" you might ask. That's an excellent question. Pilots try to touch down as slowly as possible, so from the final approach all the way through to touch down, they will gradually raise the pitch to increase the main wings' angle of attack, maintaining lift as the airplane slows to allow it to remain airborne at the slowest possible speed right before touchdown.

Most elevators have *trim tabs*—typically a small subdivision of the elevator control surface—that adjust the elevator's resting position to maintain a desired pitch and climb/descent rate for the plane's airspeed without the pilot or control system having to apply any force to the elevator. Pilots describe this as "relieving pressure" from the flight controls. Without trim tabs a pilot would have to constantly wrestle with the elevator to keep the plane flying where he or she wants it to go. Sometimes the entire horizontal stabilizer is moved upwards and downwards by fractions of a degree in order to provide trim control. This trim type is most often found on larger aircraft like long-distance commercial jets. On smaller crafts, including quadcopters, trim allows the pilot to hone in on controls and make very slight adjustments to the controls' positions.

Some horizontal stabilizers are designed to move in their entirety for elevator control, too. These combined stabilizer-elevators are aptly named "stabilators" and are much less common than the hinged elevators found on the trailing edge of fixed (or trimmable) horizontal stabilizers.

The *vertical stabilizer*—also known as the "tail fin"—is an airplane's only vertical airfoil. It controls *yaw*, the rotation of the airplane about its vertical axis and acts against *drift*, the sideways motion of an airplane while yawed. Without a vertical stabilizer the airplane might try to spin around like a merry-go-round, which would be impossible for the pilot to control. Instead, as the airplane yaws one way or another the vertical stabilizer hits the wind, causing the tail to be pushed back to its straight-flying position. This phenomenon works in exactly the same way as any other airfoil: as the airplane yaws, the angle of attack between the vertical stabilizer and the relative wind is increased, creating lift that pushes the stabilizer back to its center position. When centered, the vertical stabilizer's airfoil doesn't have an angle of attack, and its symmetrical shape allows air to flow evenly over its two surfaces. As a result of symmetry, any small amount of lift and drag force that might be created on either side is cancelled out by the other (Figure 1-18).

The only difference between the vertical stabilizer and the other airfoils on an airplane is that no opposing force like gravity is resisting the sideways lift force, making it very easy for the stabilizer to keep the airplane flying straight without any yaw. While in almost all straight flights, no yaw is best, some flight maneuvers do require yaw.

An airplane's *rudder*, located on the trailing edge of the vertical stabilizer, is the control surface for yaw. To yaw to the left, the rudder is

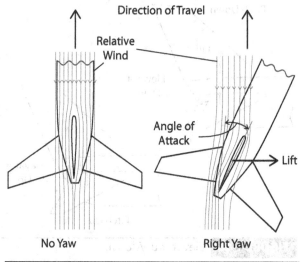

FIGURE 1-18   Right yaw.

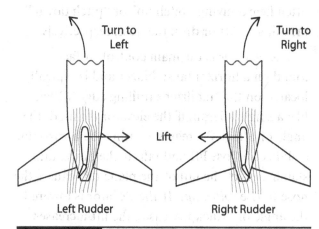

FIGURE 1-19   Rudder movement.

moved left, which increases the angle of attack to the right and also creates a lift force to the right. The airplane's tail is pushed to the right, and the nose is forced to yaw to the left. To yaw right, move the rudder to the right to create a lift force to the left, pushing the tail to the left and making the nose yaw to the right, as shown in Figure 1-19.

When taking off and landing, the airplane's nosewheel or tailwheel spends some time off the ground while the main landing gear remains on the ground. How does a pilot steer the plane to stop it from drifting off the side of the runway without control of the nosewheel or tailwheel? The rudder! That's an easy one.

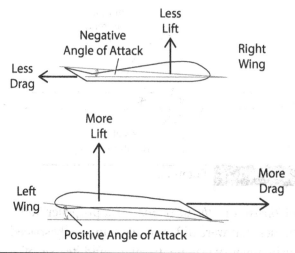

FIGURE 1-20 Angle of Attack (AoA).

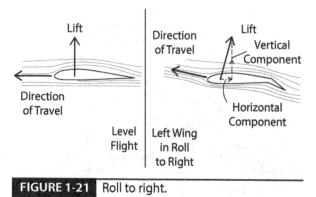

FIGURE 1-21 Roll to right.

The rudder is also an essential control surface for coordinating turns. During a right turn, the right aileron rises to decrease the angle of attack, reduce lift, and drop the right wing; while the left aileron simultaneously lowers to increase the angle of attack, enhance lift, and raise the left wing. Simple, right? Not so fast! When an airplane rolls into a turn, three negative actions occur. First, remember that more lift *always* comes with more *induced drag*. As the right aileron rises and reduces lift, the drag on the right wing also decreases and causes it to accelerate forwards; as the left wing aileron lowers and increases lift, the drag on the left wing increases and causes it to decelerate (Figure 1-20).

Second, as the aileron being moved changes the shape—or "profile"—of the wings, the *profile drag* also changes. The profile drag is lowest for very flat and thin airfoils and higher for thick and curvy airfoils. The shape of any airfoil is called its *camber*: the curvier and thicker the airfoil, the higher the camber. Since the top of an airplane's wing is already curved, the raised right aileron causes the camber to decrease; the aileron is now closer to the highest point of the airfoil. This causes the profile drag on the right wing to decrease, causing it to accelerate forwards even more. Similarly, since the bottom of an airplane's wing is flat, lowering the left

aileron will create a curve with a higher camber, increasing the profile drag and also causing the left wing to decelerate.

Third, as lift is—by definition—always perpendicular to an airfoil's relative wind, the airplane's rotation creates an additional effect. As the airplane rolls to the right and the left wing rises into the air above, its relative wind shifts from coming straight ahead to coming from slightly above. This change causes the lift force to point upwards and slightly backwards, instead of straight upwards like it does in straight-and-level flight. As a result, the left wing decelerates even further. Conversely, as the right wing descends, the relative wind shifts slightly downwards, causing the lift force to point slightly forwards and resulting in extra forward acceleration of the right wing (Figure 1-21).

These three effects of rolling combine with each other to form *adverse yaw*: the tendency of an airplane to yaw in the opposite direction of roll. The rudder can be applied in the same direction as a roll to counteract the adverse yaw (Figure 1-22).

If an aircraft needs to descend very rapidly (which can be desirable in an emergency situation that calls for the airplane to land as quickly as possible) the pilot can use a combination of ailerons (roll) and rudder (yaw) to enter a "slip." This maneuver involves yawing one side of the airplane into the wind while rolling in the opposite direction to stop the

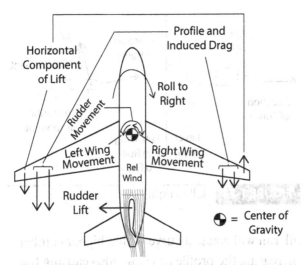

FIGURE 1-22   Top-down force diagram of aircraft in roll.

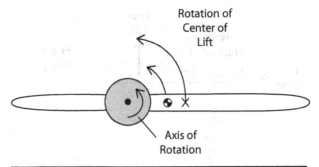

FIGURE 1-23   Rotation.

airplane from turning, and using the fuselage as a giant air brake to keep the airplane's airspeed from increasing too much as it dives to a lower altitude.

Propellers and rotors are airfoils like any other, except they rotate through the air instead of moving straight through it. This rotation makes them perfect for attaching to engines and motors, which are very efficient at creating rotary motion—much more efficient than the solenoids and hydraulic arms used to create linear motion. Feasibly, such linear actuators could be used to create wings that produce power by flapping like birds, but with today's technology they would be much less efficient than a rotary motor. Propellers come in two types: fixed pitch and variable pitch. Both types have two or more *blades*, where each blade is an individual airfoil that generates lift. In fixed pitch and variable pitch propellers alike, the principle of symmetry is the same: two or more identical blades are needed with equal spacing around the propeller's axis of rotation, simply to ensure that the generated lift is distributed evenly around that axis.

As a result, a propeller's axis of rotation is—under normal operation—also its direction

of *thrust*, or "*thrust vector.*" If a propeller had blades that were different sizes, unevenly spaced, or in any way asymmetrical around its axis of rotation, then the thrust vector would itself rotate around the axis of rotation rather than being perfectly aligned with it. This asymmetry is exactly how vibration motors work, like those used to make your phone vibrate when you get all those snaps from friends telling you how awesome your drone is! (Figure 1-23).

While a giant vibration motor on an aircraft would be unstable to say the least, it would also be incredibly inefficient at producing power. With this issue in mind, propeller manufacturers and aircraft mechanics go to great lengths to ensure that propellers are perfectly symmetrical and *balanced* out of the factory and that they remain balanced in the field, meaning that the propeller's weight is distributed evenly around the axis of rotation and that the lift produced by each blade is equal for a given pitch.

As a propeller rotates, its blades pass through the air and create a "relative wind" in the direction opposite to their rotation. Because the distance traveled through the air in a single rotation is greater at the tips of the blades than at their roots, the airspeed of the blade is also higher at the tips than at the roots, as shown in Figure 1-24.

If you double the distance from the axis of rotation, you double the airspeed. As both lift and drag increase with higher airspeed, propellers need to have a higher angle of attack

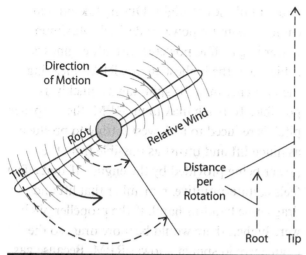

FIGURE 1-24 Front view of propeller.

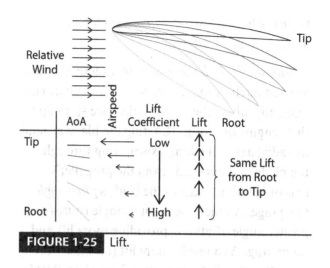

FIGURE 1-25 Lift.

at the roots and a lower angle of attack at the tips, with a gradual change in between. This variance in angle of attack from the root to the tip ensures that each propeller blade produces maximum lift and minimum drag at every point along its length.

To make this lift occur, propeller manufacturers design a *twist* into the propeller blade, offsetting the *chord* of the airfoil—the line between an airfoil's leading and trailing edges—from the direction of rotation. The closer to the root of the blade, the more twist is used, causing a greater angle between the chord and the direction of rotation. Thus the angle of attack is higher, and the lift and drag for a given airspeed are greater. Further out, the twist is smaller to create a lower angle of attack, causing less lift and less drag for that same airspeed. The amount of lift that an airfoil produces relative to its airspeed and angle of attack is called its *lift coefficient*: the more lift produced for a given airspeed and angle of attack, the higher the lift coefficient (Figure 1-25).

The twist near the root of the blade has a higher lift coefficient than the twist at the tips, which balances the lift force along the blade length when combined with the increase in airspeed from root to tip. As a result, the overall

efficiency of the propeller blade—the ratio of lift to drag—is maximized while it prevents the blade from bending due to varying lift force along its length.

Fixed pitch propellers move only one way—round and round! The amount of lift they generate (and therefore the amount of thrust and power they produce) is directly related to the propeller's speed of rotation, assuming that the air pressure, temperature, and angle of attack are constant. The speed of rotation is often measured in *RPM*, or revolutions per minute. As RPM increases, airspeed of the propeller blades also increases, producing more lift and more drag. In turn, more engine power is needed to overcome the extra drag that accompanies the higher RPMs needed to produce more thrust. This relationship means that for fixed pitch propellers, a higher engine speed always equals more thrust. Fixed pitch propellers come in different sizes and pitch angles, and matching the size and pitch angle of a propeller with the performance of the engine or motor is important. Lightweight electric motors—such as those used to power most drones—are typically most efficient at high RPM and low *torque*. Torque is the rotational force that accelerates the propeller at startup and overcomes drag at speed. Lower torque means the propeller must not have much drag, so drone propellers typically have a low *pitch angle*, each blade's average angle of attack.

Adjustable pitch propellers are exactly the same as fixed pitch propellers, except that the user can manually adjust them before they are spun up by their engine or motor. Rotating the blade around its radial axis—the long axis moving through the blade and outwards from the propeller's axis of rotation—causes the blade's *pitch angle* to change. An increase in pitch angle causes a greater angle of attack, providing more lift and more drag. As a result, more lift (and therefore more thrust) will be generated for a given RPM using a higher pitch angle than a lower one, but more power will be needed to overcome the extra drag and maintain the desired RPM.

Another way to think about pitch angle is to think of the *bite* of the propeller blades: the amount of air that the blades "bite" through each rotation. The greater the pitch angle, the more air the prop bites as it moves around. Adjustable pitch propellers usually have a mechanical coupling between the blades to ensure that each blade's pitch always remains the same as every other. If an adjustable pitch propeller lacks mechanically coupled blades, then the user must take extra care to ensure that each of the propeller's blades are set to the same pitch angle. This angle maintains the balance of lift force between the propeller blades, aligning the thrust vector with the axis of rotation and preventing the propeller from becoming a giant vibration motor! Once the pitch angle of the blades is set, an adjustable pitch propeller acts in exactly the same way as a fixed pitch propeller until it is stopped and adjusted again.

Variable pitch propellers are much like adjustable pitch propellers: the pitch angle can be changed. They are different in that they can be adjusted by the pilot or control system while the engine is moving. The pitch is increased or decreased to maximize thrust and efficiency for different phases of flight. During takeoff and climb, maximum power is desired. Maximum power in gasoline-powered aircraft engines is achieved at the highest engine RPM, meaning the propeller must be spinning as quickly as possible. To reach maximum RPM, the propeller pitch is reduced to its lowest setting to produce as much lift and thrust as possible from the power being provided by the engine. If that effect feels counterintuitive, remember that lift and drag come hand-in-hand: if the propeller pitch were higher, there would be more drag, so the engine would spin at a lower RPM. Because gas engines produce less power at lower RPMs, the speed drops even further because less power is available to overcome the drag on the propeller blades. To maximize thrust, engine RPM must be at its maximum, meaning that drag must be minimized as much as possible by lowering the pitch angle of the variable pitch prop to its minimum. However, the pitch angle shouldn't be lowered too much because some variable pitch propellers can be rotated so far that they produce *reverse thrust*!

During cruise, less thrust is needed to maintain cruise airspeed (versus accelerating during takeoff and climbing), so the pitch angle is increased to maximize the bite of the air while lowering the engine speed to a more fuel-efficient RPM. In electric motors, this speed minimizes the battery drain needed to maintain airspeed. But wait, did we not say that maximum power is reached at maximum RPM? That is exactly right. At cruise we care mostly about efficiency to maximize our range. Here's the difference between efficiency and power: *efficiency* is the amount you get of something you want (output), relative to the amount of something you consume (input) in order to get the desired output. Engine efficiency can be measured as the distance (output, in miles) that can be traveled with a specific amount of fuel (input, in gallons) or fuel needed to run for a certain period of time (output, in hours). The efficiency of gas aircraft engines is most often measured in gallons (of fuel as input) per hour (of time flying at cruise speed, which is the desired output), or *gph*. In electric drone motors, efficiency is most often measured as thrust (output, in *grams* of lifting force) for a given input of electrical power (measured in *watts*, or energy drained from the battery per second). As drone propellers are almost always fixed pitch, the propellers can be swapped out to vary pitch angle. Propellers that are larger or have a greater pitch angle produce more thrust at the expense of lower efficiency, reducing flight time in favor of better performance relative to propellers that are smaller or have lower pitch angles.

Power is the amount of energy being transferred to the propeller blades from the engine or motor. The propeller then converts power into lift and drag; the ratio of lift to drag is determined by the geometry and pitch angle of the propeller. Generally speaking, engines and electric motors are tuned to reach maximum power at their highest RPM, so it's fairly safe to assume that as RPM increases the power output also increases. If those motors or engines were not tuned, their speed would reach higher RPMs at which the power "drops off." This drop off is like running down a hill: your legs are moving quickly, but you're not using any power from your muscles to keep them moving. Compare running down a hill to running *up* a hill, which takes power from your leg muscles in order to keep moving! A very similar situation is keeping a car in first gear. You can accelerate from a standstill, but you cannot go very quickly, can you? As you accelerate in first gear, the engine RPM keep increasing. The power increases until you reach the upper end of the RPM range at which you stop accelerating. At this engine speed the car is operating above its maximum power RPM, and while your engine will be guzzling a bunch of gas, not enough power is

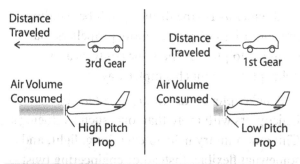

FIGURE 1-26 Distance traveled per engine rotation.

being produced to accelerate you more quickly. When you shift into second gear, you lower the engine RPM into a range that produces more power, allowing you to accelerate more quickly; third gear does the same again, etc. Changing gears in a car is similar to changing the pitch of a variable speed propeller; increasing the pitch angle is just like increasing a gear in your car (Figure 1-26).

Similarly, consider that using a fixed pitch propeller is similar to setting the car in a single gear and using only that gear for your entire journey, which limits the range of airspeeds at which you can get maximum power out of the engine or motor.

Airplanes are a great introduction to aerospace engineering, as they offer a very simple platform to understand flight. However, as the title of this book suggests, you want to learn about the wonderful world of drones! Before we jump straight into drones, you should understand another common type of aircraft: the helicopter. As you have seen before, helicopters have a large overhead rotor to generate lift.

Helicopter rotors are a little more complicated than propellers, but they use most of the same physical principles. The difference is that they are a hybrid of a propeller, a wing, and a full set of control surfaces. They rotate about an axis of rotation like a propeller, but they typically have zero or very little built-in twist. Instead, they have a fairly constant shape from root to tip.

And remember—the drones you'll be building are comprised of four or more propellers that combine to provide the same controls as a helicopter, in a much simpler way.

A helicopter's *main rotor* (or *rotors* for helicopters with more than one, such as Boeing's Chinook military helicopter) is long, light, and somewhat flexible. Instead of engineering twist into the rotor blades, their slight flexibility allows them to form their own twist as they move through the air. When in motion with a pitch angle applied, the passing air tries to twist the flexible blade back to its angle of zero pitch. Close to the root, the blade is well supported and not very flexible, so the applied pitch angle is maintained. The blade has less support and becomes more flexible the closer you move to its tip. Remember also that airspeed is higher the further out you go, meaning the quickly passing air exerts a greater twisting force on the blade's flexible tip. As they move, a natural "twist" forms in the main rotor blades through a combination of higher airspeed and greater flexibility at the tips as opposed to the roots.

Unlike airplanes, which have propellers to produce forward thrust and main wings to produce the lift needed to stay airborne, a quadcopter and helicopter's lift *and* thrust are produced by the main rotor(s). So how does a helicopter maneuver without the control surfaces found on airplanes? The answer lies in the ingenious rotor system.

As the helicopter blades rotate, the rotor control mechanism can adjust the pitch angle of each blade depending on where it is in its rotation about the helicopter body. This adjustment is called "cyclic" control because it alters the pitch angle of the blades in their cycle of rotation.

Cyclic control facilitates motion in two axes: pitch and roll. To control pitch and roll, the cyclic control mechanism moves a component called the *swashplate*—a flat bearing that changes the angle of its axis of rotation relative

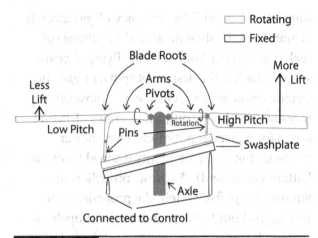

FIGURE 1-27   Swashplate.

to the axis of rotation of the rotor (Figure 1-27). The bottom of the swashplate does not rotate and is connected to the pilot's control system, while the top of the swashplate rotates at the same speed as the main rotor. A *pin or control rod* is attached to each rotor blade at one end and to the top of the swashplate at the other. Each control rod varies the pitch angle of its connected rotor blade—sometimes called the *"feathering angle"*—according to the position of the swashplate and the rotation of the blade. As the swashplate swivels away from the rotor's axis of rotation, the control rods are pushed up on the high side of the swashplate and fall down on the low side, increasing the pitch angle on the high side and decreasing the pitch angle on the low side. In turn this effect produces more lift on the high side and less left on the low side.

Pretend you're hovering in a helicopter and want to move forwards. Your only source of forward thrust is the main rotor, so you must pitch the nose down in order to apply some of the rotor's lift to generate forward thrust. Moving the cyclic forwards lowers the swashplate toward the nose of the helicopter, causing the control rods to drop as each rotor passes the low side of the swashplate at the nose of the helicopter and rise as the rotors pass over the high side of the swashplate at the tail. This movement makes the rotor pitch decrease

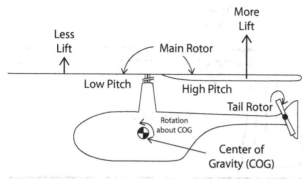

FIGURE 1-28 Side view of helicopter.

over the nose and increase over the tail, in turn reducing lift over the nose and increasing lift over the tail. The difference in lift between the front and back of the helicopter causes it to rotate about its center of gravity and pitch nose-down (Figure 1-28).

Once there, the cyclic returns to neutral, and the lift force generated by the rotor is split between forward thrust and the lift needed to counteract gravity, allowing for altitude to be maintained while moving forwards. Some of the lift that was keeping you up at altitude while hovering is now being applied to generate thrust for forward motion, so you need the rotor to generate more lift in order for the helicopter to avoid descending.

Collective control is used to alter the pitch angle of all blades at once. This alteration is much like varying the pitch angle of a variable pitch propeller: a greater pitch angle increases the blades' angle of attack, producing more lift and drag. Perhaps the simplest use of collective control is to control the rate of ascent or descent: as the collective pitch angle is increased, the blades' angle of attack increases and produces more lift and more drag. The extra lift causes the helicopter to increase its rate of ascent. Similarly, lowering the collective pitch control leads to a reduction in lift, decreasing the rate of ascent or, if low enough, allowing the helicopter to descend. In our forward-motion example, the extra lift realized by increasing the collective pitch angle allows you to maintain

the component of lift needed to oppose gravity and keep you at the same altitude while also producing the thrust needed to move forwards.

> You're probably wondering, "But what about the extra drag?"
>
> You're getting sharp! The extra drag needs to be overcome in order to maintain a constant rotor RPM and keep your extra lift, so extra engine power, along with an increase in collective pitch angle, is always needed. In addition, engine power must be decreased when the collective pitch angle is reduced. This decrease prevents the rotor from overspeeding from less drag on the blades.

And that's just the main rotor! While some helicopters have small horizontal and/or vertical stabilizers to provide stable control when they're in motion, almost all helicopters have a *tail rotor*. The only exceptions are multirotor quadcopters and helicopters designed to counteract the torque generated by the engine's power that's applied to the rotors ... more on that when you will dive into quadcopters.

A *tail rotor* is located—you guessed it—on the tail of the helicopter, with its axis of rotation perpendicular to that of the main rotor.

Flashback (or forward!) to high school physics: Newton's third law of motion states that "to every action there is always opposed an equal reaction," or in modern terms, any force applied to an object produces an equal and opposite reactive force. A helicopter's engine applies a force to its rotor blades to counteract the force of drag. As this force is applied around the main rotor's axis of rotation, the engine's force is actually applied to the rotor as *torque*, the rotational equivalent of linear force. Newton's third law of equal action and opposite reaction applies to force and torque in exactly the same way, whereby any torque exerted on one object by another is met by an equal and opposite torque.

When the helicopter is airborne, Newton's third law illustrates the fact that the torque

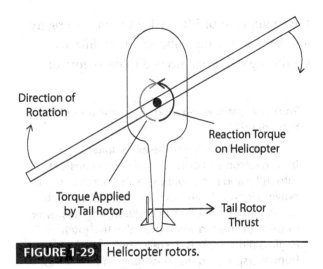

FIGURE 1-29   Helicopter rotors.

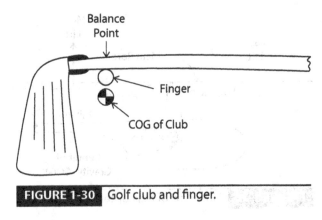

FIGURE 1-30   Golf club and finger.

applied to the rotor by the engine is met by an equal and opposite torque applied to the engine by the rotor. Since the engine is fixed to the helicopter body, this opposing torque causes the helicopter to want to rotate in the direction opposite to that of the main rotor. As a result, the tail rotor produces sideways lift to counteract the opposing torque that the rotor applies to the helicopter body (Figure 1-29).

Sideways lift is just a force like any other. Instead of causing lateral *drift* (pushing the whole thing sideways), what makes the tail rotor resist the helicopter's rotation?

The answer lies in the relationship between force and torque, along with the location of the tail rotor relative to the helicopter's center of gravity.

The *center of gravity* of anything, including helicopters, airplanes, planet earth, drones, golf clubs, people, books, buffalo, and cans of Red Bull alike, is simply the midpoint of its mass. Your center of gravity is somewhere in the middle of your torso, while a golf club's center of mass is down the shaft near the club's head. To feel a golf club's center of mass you can stick your finger out in front of you, place a golf club sideways on your finger, and find the club's "balance point"—the position along the club's shaft that your finger must support in order for

the golf club to balance there without falling off your finger. When you've found the balance point, you haven't exactly found the center of gravity (as the clubhead's shape pulls the center of gravity away from the shaft to somewhere in midair below your finger), but you're very close (Figure 1-30).

By feeling the club's weight balance either side of your finger in one dimension, you're balancing gravity-induced torques about your finger and feeling something very similar to the center of gravity, which is essentially the single balance point for every dimension. A helicopter's center of gravity is usually located on or very close to the main rotor's axis of rotation, which is usually somewhere around the middle of the helicopter body.

Now that you've got a grasp of center of gravity, you can dive a little deeper into the relationship between force and torque.

Try this exercise:

Find the nearest door and open it. How did you do it? Did you twist it at the hinge? Did you pull or push it somewhere away from the hinge? How far from the hinge did you touch the door: close in or near the edge furthest from the hinge?

When you open a door, your hand exerts a force on the door at a *radial distance* from the hinge, which is the door's axis of rotation.

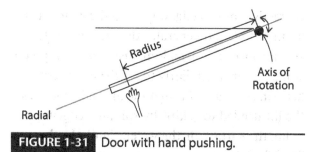

FIGURE 1-31 Door with hand pushing.

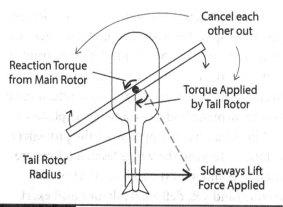

FIGURE 1-32 Top-down view of helicopter.

A *radial* is simply a line outwards from an axis of rotation that's also perpendicular to that axis, so a *radial distance* is the distance between the axis of rotation and some point along any radial. In this case, the radial is the virtual line perpendicular to the door's axis of rotation that extends out to the point at which your hand touches the door. The distance between this point on the radial and the door's axis of rotation is the radial distance, also known as the "radius," as shown in Figure 1-31.

The ease with which a door will open for a given opening force is determined by the radius at which the force is applied. Torque is a multiple of force times radius, and you can feel this by opening your door again. The further away from the hinge you push or pull the door, the greater the radius and the greater the torque; thus the easier it is to open the door. Try moving the door again, applying the same force as when you opened it. This time, apply the force close to the hinge instead of far away from it. What happens? The door will barely move, if at all. The force is the same as before, but the radius at which the force is applied is much smaller, producing much less torque on the door.

Unlike doors, helicopters don't have a hinge. Instead, their center of gravity acts as a hinge, or axis of rotation, about which the helicopter rotates. Any sideways force that is applied to the helicopter at any distance from the center of gravity will produce a torque on the helicopter.

So how does a tail rotor function? A tail rotor's blades spin with a positive pitch angle,

producing lift and a resulting thrust vector that acts sideways on the helicopter at the radial distance between the helicopter's center of gravity and the tail rotor's center of lift. Remember that a rotor or propeller's center of lift is very tightly aligned with its axis of rotation, provided that the rotor blades are properly balanced. By exerting a lift force at this radius, the tail rotor creates a torque about the helicopter's center of gravity, opposing the torque imposed on the helicopter by the main rotor blades (Figure 1-32).

When the tail rotor blades' pitch angle is controlled for zero yaw (no rotation), the torque applied by the main rotor on the helicopter body (following Newton's third law) is perfectly canceled out by the torque from the lift generated by the tail rotor. To yaw the helicopter one way or another, the pilot must simply change the tail rotor blades' pitch angle to increase or decrease their lift. In turn, the pilot increases or decreases the applied torque from that which counteracts the torque applied by the main rotors. The torque from the tail rotor will either exceed what's being applied by the main rotor, causing the helicopter to yaw in the direction of the tail rotor's lift (which is also the direction of rotation of the main rotor) or drop below it, allowing some of the main rotor torque to yaw the helicopter in the opposite direction.

Torque affects airplanes, too. As an airplane's propeller spins through the air, it also produces drag that's met by the torque delivered from the engine to the propeller. In turn, the propeller keeps Newton happy by exerting a return torque on the airplane body, causing the airplane to roll in the direction opposite of the propeller's rotation. To keep the wings level, ailerons are used to counteract the rolling effect of the torque (and yes, deflected ailerons also exert a torque about the longitudinal axis of the airplane, which in this case balances out the torque applied by the propeller).

Ailerons may be applied by a pilot or control system's input of control pressure or by using *aileron trim*. Before, you learned about elevator trim, which works in a very similar way to aileron trim. The default position of the ailerons is moved (with each aileron always moving in the opposite direction to the other) so as to maintain constant roll for a given airspeed, without the need for any control pressures to be applied. With aileron trim applied in the direction needed to counteract the torque effect of the spinning propeller, the airplane's wings remain level.

A few other forces are born of propeller and rotor rotation in helicopters, airplanes, and drones.

Gyroscopic precession is perhaps the least intuitive of the prop effects. A gyroscope is simply a spinning object—things like bike wheels, Frisbees, and skipping stones are all *gyroscopes*. On a bicycle, the gyroscopic force makes it easy for you to remain upright as the two wheels spin more quickly. Have you ever tried balancing on a bike without moving? Unless you're a very highly skilled cyclist, you will likely struggle at staying upright because no gyroscopic force is helping you balance. Similarly, if you throw a Frisbee without flicking it and making it spin, it will swivel out of control and not fly very far. The "flick" of the Frisbee turns it into a gyroscope that—when thrown

correctly—moves edgeways into the wind, acts as an airfoil, and generates the lift it needs to stay airborne and fly a long distance. A skipping stone performs similarly to a Frisbee, except it's heavier, so it falls through the air and generates the lift needed to "skip" by spinning edgeways along the surface of the water. Similarly, the "flick" of the stone causes it to rotate, turning it into a gyro that keeps it moving edgeways (flat side down) as it moves through the air and water. In fact, the main secret to getting more and more skips is to flick it as hard as possible without introducing any wobble in your release. If you can do this, the stone will rotate more quickly and maintain its gyroscopic forces for a longer period of time as the friction with the air and water tries to slow it down. The faster you release, the longer the stone will spin, giving you more and more skips!

So what does all this have to do with airplane propellers, helicopter rotors, and quadcopter propellers? Well, when they're rotating they become gyroscopes just like any other.

Try this exercise:

To understand gyroscopic forces, you should have a bike wheel in-hand (removed from the bike so you can hold it by its axle aka its axis of rotation). Hold the wheel in midair in front of you, spin up the bike wheel with your hands, and hold it with one hand at each end of the axle on either side of the wheel. Be careful, though— don't hurt yourself by letting the tire brush your nose! The rotating wheel is now a gyroscope. Try to pivot the axis of rotation by lowering one hand and raising the other. What happens? (Figure 1-33).

It will feel tough to move: this is the gyroscopic force trying to resist any movement you make to the axis of rotation—the very same force that keeps your bike upright and keeps a frisbee or skipping stone moving edgeways through the air.

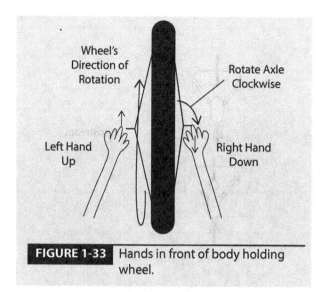

FIGURE 1-33 Hands in front of body holding wheel.

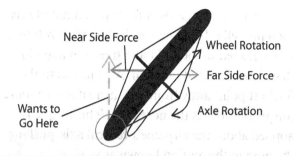

FIGURE 1-34 Bottom half of wheel, rotated clockwise.

Gyroscopic precession is a little more complicated. It might help to do this one with a long-sleeved shirt that you don't mind dirtying with some rubber, as you and your bike wheel are about to perform some magic! Once again, spin up your bike wheel as fast as you can and hold it in midair at each end of the axle. Now pivot the axle the same way you did before—drop one hand and raise the other—but this time do it really quickly, with a lot of force. You feel the resistance to moving the axle just as before, but this time the wheel also turns to the side! If you spin your wheel forwards and pivot the axle clockwise in front of you (right hand down, left hand up), the tire will turn closer to the inside of your left arm! If you pivot the axle counterclockwise (left hand down, right hand up), it will turn closer to the inside of your right arm. Magic!

Or is it magic …? What you're feeling is gyroscopic precession. When the wheel is spinning forwards, pivoting the axis of rotation clockwise (right hand down, left hand up) causes the top of the wheel to move to the right and the bottom of the wheel to move to the left. Imagine one small chunk of the tire in slow motion, passing through the lowest point of the wheel as it rotates toward you, up and around. As it passes the lowest point it is pushed—*deflected*—to the left by your clockwise movement of the

axis of rotation. Now as that same chunk of tire keeps rotating toward you and up to the closest point of the wheel to your body, it wants to remain in the left-deflected position where it was pushed at the bottom, creating a force on the wheel toward your left arm as it passes the closest point to your body (Figure 1-34).

The movement caused by that force is called gyroscopic precession, and if your wheel doesn't do this it is breaking the laws of physics. You can probably sell the wheel for a trillion dollars! Then you could build all the drones in the world!

Again you might be asking the question, "What does all this have to do with aircraft?" Think about what happens to a propeller's axis of rotation when an airplane changes pitch or yaw. Pitch the airplane up from level flight, and the axis of rotation moves from pointing forwards to pointing upwards. Do not dare try to do anything like this (or else risk chopping off your arm!), but imagine you swapped the bike wheel for a propeller as you held it in midair and spun it forwards. The airplane would be connected to the propeller out to your right-hand side, and your clockwise pivoting of the propeller's axis of rotation would be just the same thing that would happen when the airplane pitches upwards. Just like the bike wheel, as each of the propeller blades passes the bottom of its rotation, the blades are deflected to the left. As the wheel rotates 90 degrees around to the point closest to your body, it wants to maintain that left deflection and creates a torque that twists the nose of the airplane away from you. Similarly, as each propeller blade passes the top of its rotation it is deflected to the right and tries to remain in that right-deflected position. As it moves to the furthest point away from you, it creates even more torque that twists the nose away. This torque is applied about the airplane's vertical axis, pushing the nose in the motion known as yaw!

Long story short, for clockwise-spinning airplane propellers, pitching up causes the airplane to yaw to the right, and pitching down causes yaw to the left. Similarly, using rudder to yaw left causes the airplane to pitch up, and yawing right causes the airplane to pitch down.

These motions are a result of the airplane propeller's gyroscopic precession, since its axis of rotation is pivoted due to changing pitch or yaw, and elevator or rudder pressure can be used to compensate for the precession. In helicopters, the main rotor's axis of rotation is vertical instead of horizontal (as it is in airplanes), so yawing does not cause the main rotor's axis of rotation to pivot. Instead, gyroscopic precession occurs in helicopters when changing pitch or roll and can be cancelled out using opposite control pressure in roll and pitch, respectively.

The *slipstream* effect is much more intuitive. A slipstream is simply an area of altered air pressure that exists behind a moving object and is most often an area of lower pressure. Tour de France cyclists and NASCAR drivers use their opponents' slipstreams to remain in lower air pressure, causing less aerodynamic drag and allowing them to use less energy to keep pace (Figure 1-35). When their opponent becomes exhausted from powering through the higher air pressure in front of them (presuming they can't stay in someone else's slipstream), the competitor in the slipstream can use the extra conserved energy to power ahead of the opponent and—if timed correctly—win the race.

Aircraft have slipstreams just like cyclists, racecars, and any other thing that moves, but the

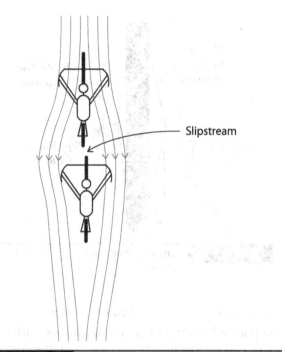

FIGURE 1-35 Cyclist in competitor's slipstream.

slipstreams of aircraft have a unique twist. Pun intended.

Remember that Newton's third law dictates that any action has an equal and opposite reaction. As a propeller or rotor rotates, the interaction of the airfoils with the air produces lift force and drag force, which are "actions" that both must follow Newton's third law. In the case of a propeller or rotor the air exerts an aerodynamic lift force on the blades, so the blades must also create an equal and opposite force on the air. This reaction to the lift force on a propeller or rotor blade is what accelerates the air through the propeller and blows it backwards into the aircraft's slipstream. By the exact same mechanism, the rotational drag force that the air exerts on the blades comes with an equal and opposite reactive force that the blades exert on the air, causing the air to rotate in the same direction as the propeller while it's simultaneously being pushed backwards by the reaction to lift force. As a result, the slipstream behind an aircraft's propeller or rotor spirals like a twister.

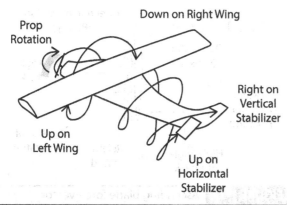

FIGURE 1-36 Diagonal view of airplane with spiraling slipstream.

In an airplane with a nose-mounted propeller, the slipstream spirals backwards and around the airplane in the same direction as the propeller's rotation. As it spirals, it hits the airplane's airfoils. This effect produces roll in the same direction as the rotation when the slipstream hits the main wings and yaw to the left as it hits the vertical stabilizer (assuming the propeller and slipstream rotate clockwise when looking toward the front of the airplane from the back) (Figure 1-36).

The effect on the horizontal stabilizer is slightly more complicated. As you might expect, the rotation of the slipstream pushes up on one side of the elevator and down on the other, causing roll in the same direction as the roll being produced by the slipstream's impact on the main wings. Remember that the slipstream doesn't just rotate; it moves backwards as well and causes the horizontal stabilizer to generate a lot of lift in addition to the roll from its rotation. If you've ever seen a tailwheel plane take off, its tail lifts off the ground almost immediately after the engine is revved up to full power at the beginning of the runway. Slipstreams are typically areas of lower pressure behind cyclists and racecars, but the air blown backwards by a spinning propeller is often called a slipstream, too. A propeller also has a low-pressure slipstream that follows its rotation, but the air

blown backwards and around the airplane is actually higher in air pressure than the ambient air. As the pilot increases the throttle to full power at the beginning of a takeoff, the high pressure and backwards movement of the air pushed on by the propeller is what causes the horizontal stabilizer to have an airspeed much higher than the airspeed at which the airplane is moving. As a result, the tail lifts off the ground right away. The same backwards motion of the propeller's slipstream also causes a higher airspeed on the vertical stabilizer. Both the rudder and elevator controls become effective at very low *ground speed*—the speed of the aircraft relative to the ground—as the aircraft begins its takeoff roll along the runway, allowing the pilot to have full control of pitch and yaw as soon as full power is reached.

Unlike the elevator and rudder, the ailerons are most often located outside of the propeller's high-airspeed spiraling slipstream. As a result, they're unable to offer any meaningful control of the airplane's roll at low ground speeds. The main landing gear takes care of this by keeping contact with the ground on either side of the airplane, preventing it from rolling until the ground speed (and therefore airspeed) is high enough for the main wings to start generating a decent amount of lift. When this airspeed is reached, the ailerons are able to offer roll control and take over roll stability from the main landing gear as the airplane's weight shifts from the landing gear to the main wings with higher and higher airspeed.

In helicopters, the main rotor's slipstream blows down and around the helicopter in the same direction as the rotor's rotation, as a result pushing on the helicopter body in that same direction of rotation. Pushing in the same direction counteracts some of the reactive torque the spinning main rotor exerts on the helicopter, and helicopter manufacturers must take this into consideration when they set the default pitch angle of the tail rotor's blades. The goal is to keep the helicopter perfectly stable in yaw without the pilot having to exert any control pressure on the pedals, which change the pitch angle of the tail rotor blades.

P-factor is one more effect of a propeller or rotor's rotation. P-factor is another force that yaws the plane, but it's a result of the angle between a propeller's thrust vector and the relative wind. The relative wind for a propeller blade has two components: the *linear component*, which is the air that comes in a straight line from the direction of travel, and the *rotary component*, which is the air the propeller pushes through as it rotates. P-factor is generated by the thrust vector's relationship with only the linear component of the relative wind.

In straight-and-level cruise flight, a minimal angle exists between the thrust vector and the relative wind. In fact, cruise airspeeds are set by airplane manufacturers to maximize engine efficiency. Return to the gyroscopic precession example of standing side-on to a propeller's rotation with the aircraft extending out to the right. Now consider that when the aircraft is flying, the oncoming air is moving—from your perspective—horizontally from left to right. Like before, increasing the pitch of the aircraft causes the propeller's axis of rotation to pivot clockwise and initiates gyroscopic precession. Once at the new, higher pitch, the propeller's axis of rotation no longer pivots, and the gyroscopic precession stops.

But now something interesting has happened! Along with the pivot of the propeller's axis of rotation, its thrust vector has pivoted, too, while the relative wind of the oncoming air continues moving from left to right, parallel to the ground. In effect, the angle between the propeller's thrust vector and the linear component of the relative wind has increased. So what?

Imagine a freeze-frame of a single *ascending* propeller blade at the point in its rotation where

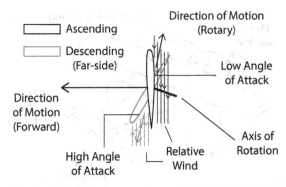

FIGURE 1-37   Ascending blade force-vector diagonal with relative wind.

it's pointing straight toward you as it moves from bottom to top of the rotation. In this freeze-frame, the blade's chord is pivoted clockwise along with the thrust vector as the airplane is pitched up, causing a change in the blade's angle of attack. Remember that angle of attack is the angle between the chord line of an airfoil—the line between its leading and trailing edge—and the direction of the relative wind. With the relative wind continuing to move in the same direction, from left to right, and the chord of the freeze-frame blade pivoted clockwise, the blade's angle of attack is decreased (Figure 1-37).

Now jump to the other side of your imaginary aircraft and look side-on to the *descending* propeller blade: this time the airplane body extends out to your left, and the relative wind of forward motion moves from right to left. Once again, freeze-frame the propeller at its closest point to you and visualize what happens to the blade's chord line if the airplane is pitched up. It rotates counterclockwise, increasing the angle between the chord line and the relative wind and increasing the blade's angle of attack.

As the ascending blade's angle of attack decreases, its lift force also decreases. Similarly, as the descending blade's angle of attack increases, its lift force also increases. In the same way that input of cyclic control pressure causes difference in lift around the rotation of a helicopter's main rotor, this difference in lift

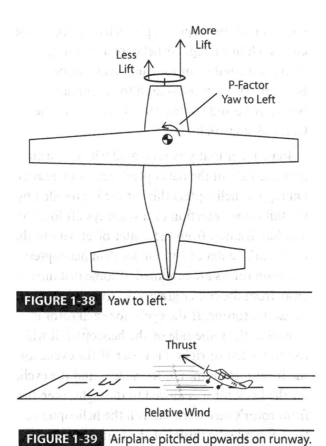

FIGURE 1-38 Yaw to left.

FIGURE 1-39 Airplane pitched upwards on runway.

force between the ascending and descending sides of the propeller causes the aircraft to yaw in the direction of the ascending side (typically on the aircraft's left side), as shown in Figure 1-38.

The resulting P-factor effect is most prevalent during takeoff when the airplane pitches up, right before it becomes airborne—a maneuver known as the "takeoff rotation." During a takeoff rotation, the nose is pitched up while the direction of travel continues straight along the runway. As a result, the relative wind is still perfectly horizontal, but the thrust vector is angled upwards (Figure 1-39).

Typically during this phase of flight, the angle between the propeller's thrust vector and the relative wind is greatest; therefore the strongest P-factor yaw is typically felt during the takeoff rotation. Pilots must use extra rudder as they rotate in order to correct for the P-factor yaw and continue traveling straight down the runway.

P-factor also affects helicopter rotors. Although it occurs during many phases of helicopter flight, P-factor is strongest when a helicopter pitches back a large amount to decelerate rapidly from high speed. Imagine standing at the side of the helicopter and mentally freeze-frame a rotor blade as it passes the point closest to you in its rotation—call this the forward-moving blade. When the helicopter is pitched back to slow down, the chord of the forward-moving blade moves with the change in pitch. As the relative wind continues coming from the direction of travel, and the blade's chord is pivoted, the angle of attack of the forward-moving blade increases, providing it with extra lift. Similarly, on the other side, the backwards-moving blade has a decreased angle of attack as the helicopter is pitched backwards to slow down, and the amount of lift on the blade is reduced.

With more lift on the forward-moving blade and less lift on the backward-moving blade, the helicopter rolls in the direction of the backward-moving blade. The helicopter pilot counteracts this roll by applying cyclic control pressure in the direction opposite of the P-factor roll.

One way to mitigate the torque effect, gyro, spiraling slipstream, and P-factor all at once is to use *counter-rotating* propellers or rotors. Just as it sounds, counter-rotation involves spinning one propeller or rotor in the opposite direction of another. Each propeller or rotor still produces all of the same effects of torque, gyroscopic precession, slipstream, and P-factor. However, as all of these forces are dependent on the direction of a propeller or rotor's rotation, two rotors moving in opposite directions to one another will produce these forces in opposing directions, causing most of them to cancel each other. Counter-rotating propellers and rotors can be positioned one above the other ("stacked") or side-by-side. However, since the complexity of the airflow in stacked configurations doesn't apply to the vast majority of drones, you'll be focusing on side-by-side.

In tandem rotor helicopters such as the Chinook, the two (and only two!) rotors cancel each other's torque, gyro, and P-factor effects.

The slipstream behaves differently. On one side of the helicopter, the blades rotate toward each other and toward the side of the helicopter body. The slipstream rotates down from the blades and pushes on the side of the helicopter, causing a sideways force on the middle of the helicopter body. On the other side of the helicopter, the blades move away from the helicopter body and apart, pushing the slipstream downwards and away from the helicopter body. Because the sideways force of the slipstream is applied on only one side of the helicopter body, it is pushed sideways—a flight condition known as *drift,* as shown in Figure 1-40.

To counteract this slipstream drift, tandem helicopters are designed to maintain a slight roll toward the side of the helicopter where the blades are moving toward the helicopter body. This slight roll causes a tilt in the rotors' lift, allocating a small amount of the lift force to counteracting the sideways force of the slipstream which in turn prevents slipstream drift. With this slipstream correction in place,

the Chinook helicopter is protected against rotor effects. Unlike a regular helicopter, it doesn't even need a tail rotor to counteract torque. But wait. A tail rotor is used to counteract both torque and control yaw, so how does the Chinook control its yaw?

Remember that yaw is created when a torque is exerted about the helicopter's center of gravity. On regular helicopters this torque is provided by the tail rotor's exertion of a sideways lift force at a radial distance from the center of gravity to the tail rotor's center of lift. On tandem helicopters, the main rotors are mounted at some distance away from the center gravity, so they can be used to exert a torque. If the cyclic for each rotor is moved to the same side of the helicopter, it will roll to the left or right. However, if the cyclic for the front rotor is moved to the left, and the cyclic for the back rotor is moved to the right, then the front rotor's thrust tries to roll the helicopter to the left while the thrust of the rear rotor tries to roll it to the right. The rolling force is canceled out, but the sideways lift force being generated by each rotor in opposite directions has to go somewhere. The *vertical component* of each rotor's lift keeps the helicopter airborne, while the *horizontal component* of each rotor's lift acts in opposite directions: the helicopter's nose is pushed to the left, and the back is pushed to the right (Figure 1-41).

Voila! Yaw control.

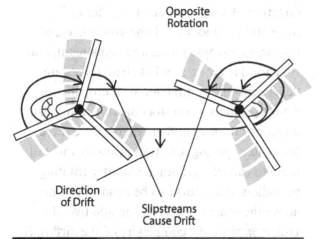

FIGURE 1-40   Top-down view of Chinook.

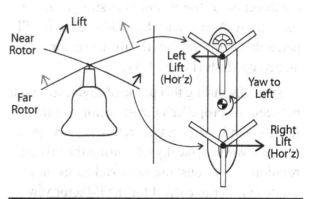

FIGURE 1-41   Front and top view of Chinook.

What about drones?

Quadcopters have counter-rotating propellers just like tandem helicopters, but – as their name gives away – they have *two pairs* of counter-rotating propellers offering incredible stability and efficiency. Each pair of quadcopter propellers that rotates in the same direction is located diagonally opposite each other on the quadcopter frame. The two pairs of propellers rotate in opposite directions, meaning two propellers rotate one way (located diagonally across from each other), and the other two propellers rotate the other way. Like tandem helicopters, the reaction torque exerted on the quadcopter frame by any one of the propellers is canceled out by another prop that creates reaction torque in the opposite direction.

Similarly, as a quadcopter maneuvers, the opposite force of each propeller cancels out spiraling slipstreams, gyroscopic precession, and P-factor from other propellers. Without any negative propeller effects, quadcopter controls are very simple, despite the fact that they lack control surfaces. All quadcopter controls are enabled when you change the relative RPM of each of the propellers.

Rate of Ascent/Descent

To increase the rate of ascent, all of the propellers must increase their speed to produce more lift, overcoming the gravitational force that acts on the mass of the entire quadcopter and allowing it to climb. Conversely, lowering the RPM of all propellers reduces the amount of lift generated relative to the force of gravity, causing the quadcopter to slow its rate of ascent or even descend (Figure 1-42).

Pitch

To pitch up the front two propellers, increase their RPM to increase lift while decreasing lift of the two rear props and reducing lift, causing the

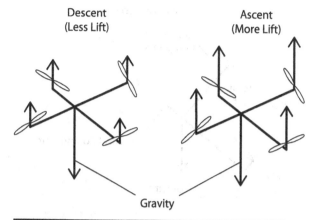

FIGURE 1-42 Diagonal view of quadcopter.

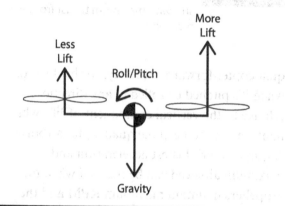

FIGURE 1-43 Side view of quadcopter pitching forward/down.

nose to lift up while the quadcopter maintains altitude. To pitch down, slow the front propellers to reduce lift and speed up the rear props to increase it (Figure 1-43).

Roll

To roll right, increase RPM on the left and decrease RPM on the right. Do the opposite to roll left.

Horizontal Airspeed

Much like helicopters, the quadcopter's acceleration and eventual horizontal airspeed is determined by the magnitude of its horizontal component of lift. Remember that the horizontal component is simply the amount of the propellers' lift force that pushes the

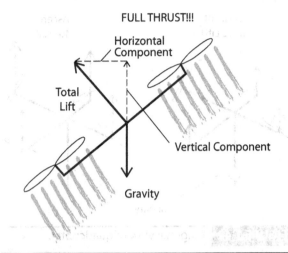

FIGURE 1-44   Side view of drone at critical attitude showing maximum thrust from each propeller.

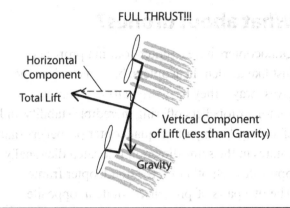

FIGURE 1-45   Side view of drone beyond critical attitude.

quadcopter forwards, sideways, or backwards when it's pitched or rolled in any direction, relative to the vertical component of lift, which fights gravity to keep the quadcopter airborne. A quadcopter's fastest acceleration and maximum airspeed can be reached when the propellers maintain maximum RPM and the quadcopter is pitched or rolled over as far as possible. Pitching or rolling the quadcopter in this way maximizes the horizontal component of the lift that the propellers are generating; that is, pitching or rolling as far as possible while still maintaining a large enough vertical component of lift to fully counteract gravity (Figure 1-44).

If pitched or rolled too far, your quadcopter may accelerate quickly horizontally, but the quadcopter's gravitational force will exceed the vertical component of lift being generated by the props, causing it to plummet to the ground (Figure 1-45).

Similarly, maximum deceleration occurs when the quadcopter is pitched or rolled to the same critical angle of maximum horizontal lift and just enough vertical lift to counteract gravity, in the direction opposite its horizontal direction of travel.

Yaw

This one is fun (and unique to aircraft with four or more rotors). Like in any other aircraft, exerting a torque about the aircraft's center of gravity controls yaw. While tandem helicopters create this torque by tilting their rotors in opposite directions to one another to produce horizontal components of lift, quadcopters can't tilt their rotors. Thankfully, they don't need to do so! A quadcopter's yaw is controlled by manipulating the reactive torque effect that each prop has on the frame—the very same effect that must be canceled by control surfaces in airplanes and helicopters. The configuration of a quadcopter's counter-rotating propellers allows it to perfectly cancel all reactive torque when it's hovering or controlling any other type of motion. To yaw one way or another, the quadcopter must do worse at canceling the torque effect of its propellers! Remember that each pair of propellers located diagonally across from each other is rotating in the same direction, and one pair rotates in the opposite direction to the other. Reactive torque is applied to the frame in the direction opposite the propeller's rotation, so the torque exerted by one diagonal pair of propellers acts in direction opposite to the torque exerted by the other diagonal pair. After all that physics, a quadcopter can move in yaw simply by increasing the speed of one

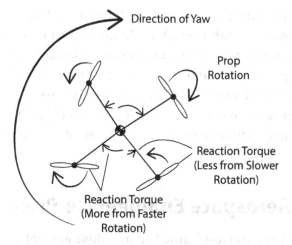

FIGURE 1-46 Top-down view of quadcopter.

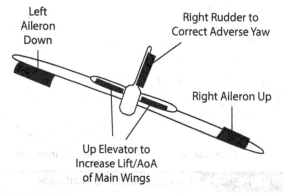

FIGURE 1-47 Rear-diagonal view of airplane in bank.

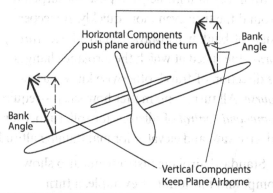

FIGURE 1-48 Rear view of airplane in bank showing total lift and horizontal/vertical components.

diagonal pair of propellers relative to the other (Figure 1-46).

Like with pitch and roll control, if one pair of rotors increases its RPM then the other pair must lower its RPM. This balance ensures that the total lift force of all propellers combined remains constant and allows the quadcopter to maintain its altitude (or rate of ascent/descent) while moving in yaw, pitch, or roll.

Drift

On an airplane, the vertical stabilizer prevents sideways drift. A helicopter simply tilts the cyclic in the direction opposite to the drift, and this action increases the horizontal component of lift from the main rotors to stop the drift. A quadcopter doesn't have a vertical stabilizer, and its propellers don't have cyclic control. Instead, a quadcopter must adjust its pitch and/or roll to create the horizontal component of lift needed to oppose any force, like that of any wind, which may be causing sideways drift.

Airplanes, helicopters, tandem copters, and quadcopters are the same as all other types of aircraft in that their controls can be used independently or can be used together. *Compound control* is when more than one axis of control is used at the same time.

For an airplane to turn, it uses its ailerons to roll into the turn and rudder to correct for adverse yaw. When rolled over, the horizontal component of the lift provided by the main wings increases, and the vertical component of lift correspondingly decreases while the total lift produced by the wings remains constant. To restore the lost vertical lift, the pilot or control system must raise the elevator to increase the main wings' angle of attack and the total lift force produced (Figure 1-47).

The extra lift force is split between its horizontal and vertical components in amounts proportional to the angle of *bank* (the amount of roll) (Figure 1-48).

The extra vertical lift restores what has been lost to the horizontal component, which had

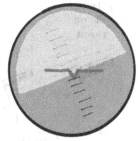

Turn Coordinator   Attitude Indicator

FIGURE 1-49 Turn coordinator and attitude indicator.

been increased by rolling into the turn, while the extra horizontal lift helps to push the airplane around the turn even more quickly. A steeper angle of bank will always lead to a faster *rate of turn*—the speed at which the airplane changes its direction of travel, otherwise known as its *course*. All turns—no matter how steep—require *compound control* of ailerons for roll, rudder for adverse yaw, and elevator for maintaining altitude.

Standard airplane instruments also show compound motion. For example, a turn coordinator shows roll relative to gravity and yaw relative to forward flight, while an attitude indicator shows both roll and pitch relative to the horizon, as shown in Figure 1-49.

For a tandem copter to yaw while simultaneously ascending, it must create opposite tilt of each rotor with the cyclic. At the same time, it must also increase lift with the collective and increasing engine power to overcome the extra drag that comes with increased lift. This yaw is another example of compound control.

Quadcopters have two different types of compound control: acrobatic mode and stabilization mode. In acrobatic mode, the pilot can move multiple control axes at once to produce compound control. In the quadcopters you'll fly and build, you will use stabilization mode. Stabilization mode uses the flight computer to automatically try to return the quadcopter to a hovering position parallel to the ground. But what does that mean in actual practice? It means that

when you pitch forwards and yaw, the stabilizing mode outputs controls to the quadcopter to make it rotate as if the yaw axis is perpendicular to the ground (instead of perpendicular to the pitched forward quadcopter). Later in the book you'll practice flying in stabilization mode, and you'll quickly understand the implications of this easier-to-fly setting.

Aerospace Engineer De-Brief

Wow! You just learned an incredible amount about the basics of flight: how basic airplanes, helicopters, and, most importantly, drones stay airborne and the different control mechanisms for different vehicles.

Over the last decade, drones have evolved from a niche aircraft primarily used by the military to the most popular category of consumer electronics available today. But why did this happen so quickly? Believe it or not, it boils down to the fact that the parts used to build an effective drone used to be either way too expensive for the average consumer to purchase, or just not yet invented until recently. Despite having fewer moving parts, drones still need complex computers.

To prove this point, you should know the most essential components of an advanced consumer drone: the camera, the sensors, and the controller.

While modern drones can perform a variety of features (some of which you'll learn later), the majority of consumer drones are primarily used to capture aerial photos and videos. Hobbyists use their drones to capture footage of buildings, nature, and even people. To capture the best footage possible, most drones are equipped with built-in, high-quality 4K video cameras. 4K video was first made available to consumers about a decade ago and is four times the resolution and quality of regular HD video. But the technology was so new that the cost was way too expensive for even most professional videographers to afford—when the first camera came out, it cost $3000 a day just to rent it! But fast-forward 10 years, and the price of 4K video cameras has plummeted to just a few hundred dollars. This drop in price means that even budget-friendly drones can capture great quality video, which has only incentivized more hobbyists to quickly jump into the industry.

While a camera is often the most discussed part of a drone, it technically isn't an *essential* component. However, sensors like a GPS chip are necessary if you want to build or buy an advanced, self-stabilizing drone. GPS stands for Global Positioning System and is a physical chip that can calculate exactly where your drone is by communicating with satellites in space. Drones use this technology to hover in place, self-navigate, and avoid obstacles. Essentially, GPS provides the technology that makes drones easy and fun to fly, which is a huge reason why so many hobbyists have quickly fallen in love with flying. But similar to 4K video, GPS technology had been cost-prohibitive until about 15 years ago, when it started being integrated into consumer electronic devices. Though you can now find a GPS in almost every consumer electronics device on the market, a GPS navigation chip in your car, let alone your drone, was uncommon!

Cell phone are also an important technology for drones. Most advanced consumer drones now connect to your cell phone while in flight, and they transmit things like speed, altitude, and even live video. While most drones still rely on a physical hardware controller to relay essential movements, they wouldn't be able to fly without the phone acting as a reliable companion. But how exactly do they communicate? Most drone manufacturers offer an accompanying iPhone or Android app that users can download and link to the device. But mobile apps were only invented about seven years ago, and

FIGURE 1-50 DJI Phantom 4, a ready-to-fly consumer drone.

smartphones were created not long before that. Before this, drones could only be controlled by a remote control and wouldn't have benefited from the features that are now provided by a phone.

These three technologies are just three examples of recent technological advancements that have made modern consumer drones possible. Almost every piece of hardware inside a drone is either a relatively new invention or has benefited from a massive price drop as technology has enabled mass manufacturing of the part. So next time you're flying a drone, just imagine how lucky you are to live in a year when you can take things like 4K cameras, GPS, and cell phones for granted (Figure 1-50).

Drones don't just appeal hobbyists. Over the last decade, more and more businesses have started exploring how drones can help them operate more quickly and efficiently, all while saving them money. While drones may eventually help businesses in all industries, a few types of companies will particularly benefit from the use of drones. To dive deeper into how drones can help businesses, you should know three of the most promising commercial use cases for drones: transportation and delivery, remote surveying, and cinematography.

Imagine stepping outside and seeing a drone hovering over your front yard, lowering itself to deliver a pizza or bag of groceries. While these scenarios sound outlandish, they aren't that far-fetched. In fact, major companies like Amazon and Domino's are already developing drones to make this a reality. In 2013, Domino's used a drone to deliver a pizza in the United Kingdom. The company hired a UK-based drone company to develop a specialized drone that would carry a pizza payload, and they hired an experienced pilot to fly it four miles across town. While the test was more than likely a publicity stunt, it showed the potential of how businesses could use drones to deliver food directly to your doorstep.

While drone pizza deliveries may still be a few years away, Amazon thinks they may be able to start using drones to deliver your online shopping orders very soon. The company has research labs in the United States, the United Kingdom, Austria, and Israel, where engineers have built over a dozen different prototypes in their mission to discover how to best use drones to deliver packages. The project is called Amazon Prime Air, and the company says it will soon be able to deliver packages under five pounds in "30 minutes or less." The drones will fly below 400 feet and be equipped with advanced "sense and avoid" technology so they can fly themselves to their destinations. When will you be able to try Prime Air? Amazon says that they will launch the project when they have the regulatory support to legally and safely operate the drones in different countries around the world.

In the future, drones will also automate dangerous and tedious surveying jobs. One of the main benefits of a modern drone with a high-quality camera is that it can capture video footage much more quickly and safely than a human. For example, a construction crew preparing to renovate a bridge may need comprehensive 360-degree photos and videos of the bridge's every nook and cranny. Before the advent of drones, a construction worker

had to strap on a harness and dangerously scale the bridge for a physical inspection or hire an expensive aerial helicopter to take the photographs. Now, a remote operator can capture high-quality footage from a safe distance for a fraction of the cost.

The final industry in the midst of being revolutionized by drones is the cinematography industry. Every day, filmmakers and journalists around the world rely on helicopters to capture aerial shots of anything from important news events to that awesome chase scene in your favorite action movie. But chartering helicopters is expensive; some fleets can run thousands of dollars an hour once you factor in the cost of a film crew, equipment, and pilots. The result is that aerial shots typically only exist in high-budget films and are almost never seen in indie or amateur video, which is a shame because high-quality aerial cinematography can add an entirely new perspective to films. Luckily, drones have stepped in as affordable replacements, and more and more film crews are benefiting every day from the low-cost, high-quality footage that can be captured by drones.

But are film crews just using your average consumer drone to capture footage for films? Not exactly. Film crews often use drones that can carry a professional, DSLR camera. These drones are often called "flying platforms" because instead of having many built-in components and features like a consumer drone, they are bare-bones platforms designed to be used for one purpose, which in many cases is filmmaking. Interestingly, they often contain anywhere from six to eight propellers, since they need to be more stable and carry more weight than your average drone.

While these platforms are more expensive than consumer drones, they are still much more affordable than constantly chartering a helicopter to capture footage. That being said, even affordable consumer drones have cameras that are more than effective enough for an amateur photographer to capture aerial footage. In fact, many hobbyists have used consumer drones to capture high-quality footage of events like a kid's sporting event or a friend's wedding. If you own a consumer drone and want to start capturing and sharing footage from the air, your drone should be more than good enough to get started.

Outside of the film industry, journalists are also benefitting from the affordability of drones. While news stations in major cities like New York and San Francisco may have their own full-time helicopter crew to capture footage of news events, smaller cities and publications are often unable to give their viewers a live look at news happening in their city. But this lack of live footage is now changing; anyone with a drone can either capture aerial footage of a news event or stream it live to viewers online. Imagine a newspaper reporter showing up on the scene of an accident and instead of pulling out a pad of paper, he or she pulls out a pocket-sized drone to capture overhead shots of the scene. As drones become even more popular, expect to see more and more journalists integrating the flying crafts into their toolsets.

While drone technology has now evolved enough for the aircrafts to benefit businesses in almost all industries, the US government isn't yet ready to let just anyone and everyone fill the skies with commercial drones. Specifically, the Federal Aviation Administration (FAA) maintains a list of rules governing the operation of commercial drones. These rules include specifications like the drone must remain in sight of the operator at all times, and drones can only be flown during the daytime below 400 feet. Additionally, operators of commercial drones must be at least 16 years old and have a remote pilot certificate, or be directly supervised by someone who has a certificate. To obtain a remote pilot certificate, operators will have to pass an aeronautical knowledge test at an FAA-approved testing center. Even though these restrictions are only for operators flying drones for commercial purposes, it's important to be aware of the rules in case your hobby ever turns into a part-time job or even a career in the drone industry.

PROJECT 2

Wing-Free Flight

Now that you've narrowed your focus to a specific drone category, you'll learn about the six basic components that make up a quadcopter.

The drones you'll make contain six key components (Figure 2-1):

1. Frame
2. Powertrain
3. Power Supply
4. Power Distribution
5. Radio Control
6. Flight Computer

1: Frame

The drone's is its core support structure. All components will attach to the frame. Since frames

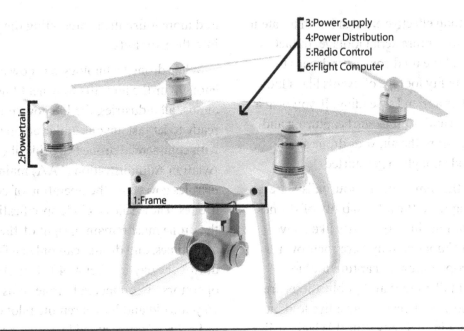

3:Power Supply
4:Power Distribution
5:Radio Control
6:Flight Computer

2:Powertrain

1:Frame

FIGURE 2-1   Six key components.

need to be strong enough to withstand the stresses of flying while carrying payloads, they are usually made of either fiberglass-infused plastic or carbon fiber. Some people have made drones out of traditional construction material like aluminum, and others have made drones out of slightly less traditional materials such as taxidermy cats. Really, anything works as long as the frame is strong enough to support the other five key components. Let your imagination go wild!

Some frames come with additional accessories such as landing gear and propeller guards. While these additional structures are useful, they do tend to add additional weight to the drone. And remember, extra weight means less flight time.

Frames have "arms" that extend outward from the center. Each arm has a powertrain unit attached, which includes the spinning propeller.

A bit of nomenclature (Figure 2-2):

Frames with 4 arms are called "quadcopters."

Frames with 6 arms are called "hexacopters."

Frames with 8 arms are called "octacopters."

FIGURE 2-2   Quadcopter frame.

2: Powertrain

The powertrain is responsible for getting a drone off the ground. The powertrain itself consists of three critical components: propeller, brushless motor, and electronic speed controller. For power, every propeller requires a motor, which requires an electronic speed controller to set its speed. If you have four arms, you have four propellers, and you need four brushless motors and four electronic speed controllers.

Propellers generate lift. Propellers are connected directly to the brushless motor shafts, spinning them at very high rates of speed to lift the multirotor into the air.

As discussed in the previous section about propellers, the faster a fixed-pitch propeller spins, the more lift it generates. To achieve a high rate of rotation, quadcopters need to utilize lightweight high-performance electric motors.

Brushless motors are motors capable of spinning at high rates of speed. These motors are connected directly to the propellers, therefore generating lift. Since multirotor craft self-balance, they need to be able to change each propeller's speed instantaneously to generate more or less lift to compensate for the ever-changing environmental conditions.

Electronic speed controllers (ESCs) tell each individual brushless motor how fast to spin. ESCs have two inputs and one output. They take in power directly from the battery (usually 12 volts). They also take in a pulse width modulation channel, which is the way the flight computer communicates with the ESC and tells the ESC how fast to spin the brushless motor. And you guessed it—the only output of an ESC is the connection to the brushless motor. And as you know, the brushless motor is connected to the propeller (Figure 2-3).

3: Power Supply

Unlike airplanes and helicopters, almost all multirotor aircraft run on electric power. With top-of-the-line lithium polymer (LiPo) batteries, the drones can stay airborne for 10 to 30 minutes at a time. Since flight time is relatively short compared to other types of aircraft, operators tend to be very

FIGURE 2-3 | Powertrain components.

mindful of the weight the craft is carrying. The more weight, the more energy required to lift the aircraft and the less time the battery will last.

LiPo batteries typically power drones. These are extremely high-performance batteries that can deliver massive amounts of amperage very quickly. In basic terms, that means these batteries are able to quickly spin up the brushless motors to get your drone airborne! These are the same types of batteries used in electric cars like the Tesla Model S, X, and Three.

LiPo batteries have two main characteristics: the number of cells and the capacity. Most batteries used for drones run at 11.1 volts. As the battery loses its charge, the voltage drops. Drones typically opt for three-cell batteries with varying capacity. The higher the capacity, the longer your drone can stay airborne. On the other hand, the higher the capacity, the larger and heavier the battery is. As you've begun to noticed, it is always a tradeoff with weight versus performance when it comes to multirotor aircraft.

These batteries tend to use a XT-60 plug, which is a standard plug for radio-controlled devices (planes, helicopters, and drones). They also have a "monitoring" cable for use during charging. When charging these

FIGURE 2-4   LiPo battery.

FIGURE 2-5   LiPo charger.

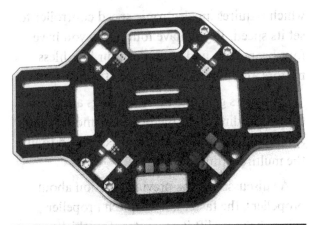

FIGURE 2-6   Power distribution board.

high-performance batteries, it is always critical to use a reliable charger (don't worry: you'll learn how to choose the perfect charger later in this book) and stay in vicinity of the battery. If something goes wrong, the battery will bulge, smoke, then burst into flames. Remember: safety first! (Figures 2-4 and 2-5).

4: Power Distribution

Having a high-performance battery is useless unless it's hooked up to the components of the drone. The power distribution routes the battery power to all components powered by a 12-volt input source. On the drones you'll be building, the power distribution includes the ESCs and additional accessories such as lighting and landing gear. Note that flight computers are typically powered by a lower voltage (three volts or five volts). Luckily, some ESCs step down the voltage and distribute the power to the flight computer via the PWM cables. Other drones will have a separate voltage converter instead of utilizing the ESCs for this purpose. Either way, as long as the flight computer gets powered, your drone will be ready to fly (Figure 2-6).

5: Radio Control

For the majority of this book, you'll focus on drones flown by an operator via radio control. There are ways to preprogram a drone's route, but the simplest way to get started is with a radio transmitter and receiver. Radios come in a variety of shapes and forms, but basic starter packages will have two critical components: a transmitter and a receiver.

The transmitter is the remote control you've probably seen before. It has two joysticks and a series of buttons as shown in Figure 2-7. Transmitters have a different number of channels. Each channel represents one control input. Most drones require at least four inputs, but each additional input gives your drone more capabilities such as changing the flying mode, deploying landing gear, turning on lights, etc. We recommend a controller with six inputs or more.

The receiver is a small device that will be attached to your drone's frame. It connects directly to and is powered by the flight computer.

FIGURE 2-7 Radio transmitter.

FIGURE 2-8 Flight computer.

The receiver will come preprogrammed and already synced to your transmitter. It will relay the inputs on the transmitter to the flight computer.

6: Flight Computer

The flight computer is arguably the most important component of the drone. Think of it as the multirotor's brain. It controls everything—from turning lights on and off to balancing the drone.

Flight computers have several inputs and outputs. On the input side, there are radio receiver inputs and sensor inputs.

Radio receiver inputs relay, in real time, the current action being taken on the radio controller by the drone pilot. Flight computers come equipped with a variety of sensors—usually an accelerometer and gyroscope. Accelerometers measure force movement in all directions. Gyroscopes measure and maintain rotational motion.

Flight computers output signals to PWM cables that are connected to the ESCs. These ESCs then control the motors' speed.

By using a combination of the two sensors, the drone is able to figure out if it is flying level. By further integrating the radio receiver input, the flight computer is able to change the speed, altitude, direction, and more of the drone.

The flight computer is constantly recalculating all of its inputs and sending live changes to the PWM cables. What does that mean for your drone? The sensors that the flight computer uses to recalculate its PWM output, which change the speed of each of the propellers, instantly acknowledge disruptions like heavy wind or other obstacles; they instantly restabilize the aircraft.

Some flight computers have even more inputs, such as GPS transponders, which let the flight computer know the drone's exact position in space. Using GPS, the drone can not only stay level, but can fly on its own to a new location and stay at that specified location until further input is received (Figure 2-8).

PROJECT 3
Ready for Takeoff

Now that you understand the parts of the drone, you'll learn how the drone itself actually flies!

All aircraft operate with reference to three key directions of motion known as the Aircraft Principal Axes. These three axes are pitch, roll, and yaw (Figure 3-1), which you already learned about in Project One, but you'll be reminded about them here.

Pitch

Pitch is the rotation of the airplane about its lateral axis—the axis that passes through the side of the airplane at right angles to the fuselage. Pilots are often heard saying "pitch up" or "pitch down," which mean to lift or drop the nose, respectively.

On a quadcopter, pitching the front up or down moves the quadcopter forwards or backwards.

Roll

Roll is the rotation about the longitudinal axis—the axis that goes through the middle of the *fuselage*, or body, of the aircraft.

On a quadcopter, rolling to the right or left lets the aircraft move to the right or left without moving forwards.

Yaw

Yaw is the rotation along the vertical axis—the axis that goes through the center of the fuselage, or body, of the aircraft.

On a quadcopter, yawing clockwise or counterclockwise lets the aircraft rotate or spin.

Drones use these same principal axes (Figure 3-2). Drones fly forwards when their pitch tilts them forwards. Drones fly backwards when their pitch tilts them backwards. And so on and so on. Here is the really cool part: the flight

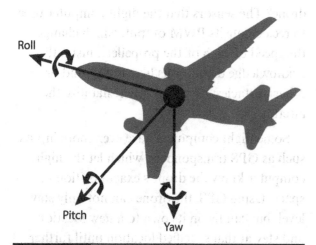

FIGURE 3-1   Principal axes on an airplane.

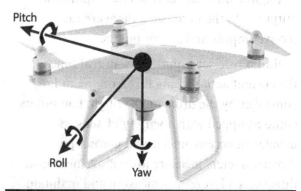

FIGURE 3-2   Principal axes on a drone.

computer calculates these axes in real time and adjusts the propellers as needed to move the quadcopter exactly as the operator desires.

When you first fly a drone, you'll begin by focusing on just two of these axes: pitch and roll.

As you become more comfortable with basic movement, you'll begin to incorporate yaw into your control, letting you operate your multirotor craft in a more fluid manner.

Safety First

You'll review FAA compliance and educate yourself on the most responsible way to enjoy our new favorite hobby.

Before you can actually start flying, you need to become a certified drone pilot. Luckily, the process is simple and can be done entirely online. Before you dive into the registration details, you should know the rules that govern flying your multirotor aircraft. Keep in mind, these rules apply to the United States only. For countries outside the United States, consult your local government for drone-related laws. These rules and regulations change over time, so please always check with the FAA for the most up-to-date rules on drones. As of 2017, the FAA's drone registration, called Small Unmanned Aircraft System (sUAS), costs $5.

The FAA regulates all drone flights in the United States. Since the drone industry has grown so rapidly over recent years, the FAA has divided drone operations into either hobbyist or commercial flights. Currently, if you're a hobbyist drone operator flying purely for recreation, you're free to fly without FAA certification. That being said there are still certain rules to abide by, which you'll learn later in the book. And if you're a commercial

drone operator (even if you're using your personal hobby drone for commercial use) you're required to either obtain a remote pilots license or fly under the direct supervision of someone that has a permit for performing commercial operations. It's important to remember that the FAA is a federal agency, tasked with keeping America's skies safe and regulated. As a result, they take violations seriously and can assign civil or criminal penalties for operating your drone in a way that breaks FAA regulations.

"Do I Need sUAS Registration?"

Not necessarily! If your drone weighs less than 250 grams on takeoff, or you only fly your drone indoors, you don't need to register your drone. In all other cases, however, registration is required. And it's very important to register. In addition to the obvious safety concerns, the FAA can impose a civil fine of up $27,500.

"What Do I Need to Do Once I've Registered?"

Not very much. You should always keep your drone registration card and number on hand. The registration process, which you'll discover in the next chapter, provides you with a ready-to-print card. In addition, it is extremely important to attach your registration number to your drone. Using a Sharpie pen is typically the easiest way.

"What Am I Allowed to Do?"

For up-to-date rules, check with the FAA. As of 2016, the FAA's Safety Guidance includes the following (Figure 4-1):

- Fly below 400 feet
- Fly within visual line of sight
- Be aware of airspace requirements
- Do not fly over people
- Do not fly over stadiums or sport events
- Do not fly near emergency response efforts
- Do not fly near aircraft or near airports
- Do not fly under the influence

FIGURE 4-1   FAA regulation basics.

Beyond all the rules and regulations, it is critical to keep in mind that drones are indeed dangerous. The propellers spin at extremely high rates and can cause injury or even death. The drones can weigh several pounds, and if the drone falls onto or hits a person or animal at any speed, serious injury will occur. Always use your best judgment and err on the side of caution. Drones are still a new hobby and industry, so it's important to show regulators that you're able to enjoy your hobby in a responsible and safe manner.

PROJECT 5

License to Drone

You're now ready to learn the FAA licensing and registration process, which will allow you to fly drones of nearly any size!

The actual FAA registration is extremely easy. You need to register once and pay a $5 fee. The single registration applies to all of your drones.

To register, go to **diydronebook.com** and find the link to the FAA's sUAS registration system.

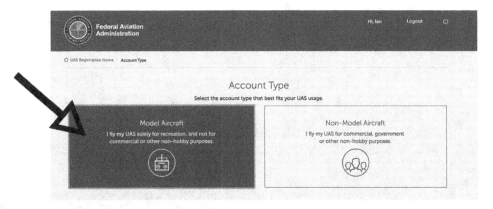

FIGURE 5-1 sUAS registration website.

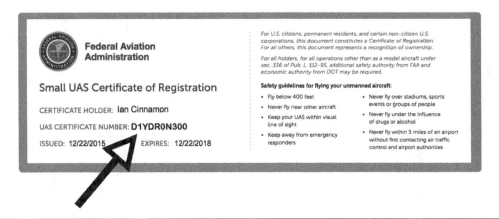

FIGURE 5-2 Sample registration certificate.

Once on the FAA registration website, select "Model Aircraft" as shown in Figure 5-1.

Once you fill out your personal information, you will be prompted to pay the $5 registration fee. After paying, you will be presented with your Registration Certificate as shown in Figure 5-2.

At this point, the final step is to attach your registration number to your drones. You'll discover a few labeling options during the later chapters of this book.

Congratulations! You're now a registered Drone Owner!

In the next section, you'll get your hands on your first drone and learn how to fly.

SECTION TWO
Small-Scale Fun

Using an off-the-shelf beginner drone available for $15, you'll become an expert drone flyer, working up to flips and high-speed maneuvers.

Your First Drone

In this project, you'll explore your new palm-sized drone. You'll learn about the different components and familiarize yourself with the basic controls.

So far, you've done a lot of reading. You learned about the basics of flight and now understand the components of quadcopters. You have even registered with the FAA. Now it's time to fly your first drone. The model drone for this chapter is the Cheerson CX-10 (Figure 6-1), which is the perfect drone for first-time fliers . The drone is self-balancing, fits in the palm of your hand, is perfect for indoor flying, and costs only $15.

To purchase the drone, go to **diydronebook.com**.

If you're feeling ambitious, the slightly more expensive Cheerson CX-10C drone includes an integrated camera, as shown in Figure 6-2. While the camera is low resolution, it's still fun. The CX-10C is available for $25. Find a link to the product at **diydronebook.com**.

Now that you're holding your first drone, take a close look at it and make sure you understand all of its components. In the box, you should see the quadcopter itself, the controller, and the charging cable. Put batteries in the controller and charge up the quadcopter.

While the quadcopter is charging, examine the pieces. The drone clearly has four arms, four motors, and four propellers as shown in Figure 6-3.

When turned on, the drone has a couple of LEDs to help you know its direction of flight. Blue is the front; red is the back (Figure 6-4).

The plastic shell over the middle of the drone contains all of the other components: the battery, flight computer, ESCs, and so on as shown in Figure 6-5.

Now look at the controller. You'll notice two joysticks, switches next to each joystick, an on/off

FIGURE 6-1 Cheerson CX-10 drone.

Camera Module

FIGURE 6-2 Cheerson CX-10 camera drone.

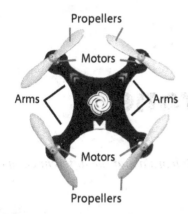

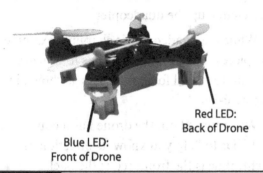

FIGURE 6-3   The four arms of the quadcopter.

FIGURE 6-5   The other components with the plastic shell removed.

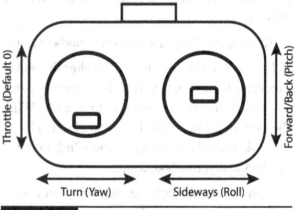

FIGURE 6-4   LEDs to help you orient while flying.

toggle, and a couple of buttons you can activate by pressing firmly down on each joystick.

The two joysticks are your primary control mechanisms. You'll learn to fly in "Mode 2," which is the default flying mode for most quadcopters (Figure 6-6). In Mode 2, the controller is set up in the following way:

- Left Joystick Up/Down: Throttle. This control powers up the propellers and controls the altitude.

- Left Joystick Right/Left: Turn (Yaw). This control allows the aircraft to rotate clockwise and counterclockwise.

- Right Joystick Up/Down: Forward and Backward (Pitch). This control makes the drone tilt forwards and backwards, generating motion in those directions.

FIGURE 6-6   Mode 2 control.

- Right Joystick Right/Left: Sideways Motion (Roll). This control makes the drone tilt left and right, generating side-to-side motion.

The small switches next to the joysticks control the "trim." If your drone tends to drift in a certain direction even when the joysticks are not being pushed, you can tap these switches to programmatically slightly shift the drone in those directions.

The controller has two more buttons you can actuate by pushing downward on the left and the right joysticks. In general, these are fully programmable and can do anything from turning on LED lights to taking photos. On this Cheerson drone, the buttons allow the drone to do complete flips. You'll practice flipping the drone in later chapters.

PROJECT 7

Takeoff

It's time for takeoff! You'll practice rapidly ascending and controlling the craft during the early stages of flight.

Enough theory: time to get flying! In this project, you're going to practice lifting the drone off of a table, hovering it, and setting it back down. You'll also trim the drone and prepare it for more exciting flights in the next few projects.

The Cheerson is small enough to fly inside. In fact, because the drone is so small and light, it's much better to fly indoors; any moderate wind would sweep the drone out of the sky. Find a prime indoor location: an empty living room, a flat bed, or a kitchen table. The drone is much safer than ones you'll be flying later because its propellers are made of soft plastic that might sting if they hit you but won't cause any serious harm. However, always use caution while flying and never fly the drone near anyone's face.

Turn the drone on (there's a switch on the back). You should see the red and blue lights turn on. Position your drone on the center of your landing pad (most likely a piece of furniture). Be sure the landing pad is flat.

Now, turn on the controller. You should see a light on the controller turn on. Make sure the controller is zeroed (Figure 7-1). In Mode 2, that means:

FIGURE 7-1   Zeroed Mode 2 controller.

- Left joystick: Joystick should be oriented to the middle bottom.

- Right joystick: Joystick should be oriented in the dead center.

Before the drone will fly, it needs to be armed. When you build your own drone from scratch in the later chapters, you'll choose how you want to arm your drone. The Cheerson, along with other ready-to-fly drones, comes with a very simple arming mechanism. Simply take the left joystick and move it all the way up and then all the way back down, as shown

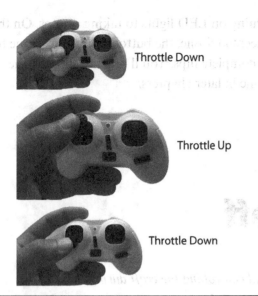

Throttle Down

Throttle Up

Throttle Down

FIGURE 7-2   Arming the drone.

Trim Switches

FIGURE 7-3   Trimming the drone.

in Figure 7-2. The motors will not spin, but you should hear a series of beeps. You're armed and ready to fly! Once armed, you have 60 seconds to start flying before the drone disarms itself. If the drone disarms itself, you need to follow the arming procedure again before flying. If you try to take off, and the motors do not spin, repeat the arming procedure.

It's time for takeoff! Slowly move the left joystick up while keeping it centered in the left/right directions. You should see the drone slowly lift up. As the drone lifts up, you will need to

constantly adjust the left joystick (the throttle) to control the altitude.

Keep the right joystick centered (you can remove your right thumb from this control).

Watch your drone. Is it spinning clockwise or counterclockwise? Is it moving to the right or left? Forward or backwards? If so, tap on the trim buttons until your drone stays perfectly centered (Figure 7-3).

Now slowly let back down on the throttle and bring your drone in for a gentle landing.

Congratulations! You just completed your very first drone flight!

PROJECT 8

Basic Movement

Now that your drone is airborne, you'll practice forward, backward, and sideways motions. We'll also learn how to fly the drone around obstacles.

You know how to take off (and possibly land or crash-land) so it's time to really get flying. First,

charge your drone. It is always best to fly with a full battery.

Go back to your drone's landing pad/bed/table/flat-open-area. Turn your drone on, arm it, and get it airborne. Since you trimmed your drone in the last section, it should remain relatively stationary in the air.

With your left thumb, continue to move the left joystick up and down to control the drone's altitude. You'll notice that you need to gradually increase the throttle to maintain the same altitude—since the battery drains over time, the throttle must be increased to counteract the lower battery voltage (and therefore slower motors).

Now that the drone is airborne again, use your right thumb to move the joystick up, down, left, and right. Your drone should instantly respond to your commands. Set up some obstacles and start flying your drone around. Try to get the hang of maneuvering the quadcopter around tight corners while maintaining the same altitude.

If you imagine the drone remaining at a constant altitude and simply moving in two dimensions (forward, backward, right, and left), the drone is simple to fly. Unlike a car or airplane, the drone is able to instantly change direction without any forward momentum. Cars use wheels to guide turns, while drones can simply alter the pitch or roll to change directions or suddenly stop (Figure 8-1). This increased maneuverability is part of what makes quadcopters so unique. They're able to navigate around tight spaces, hover in place, and perform a variety of other activities standard aircraft simply cannot perform (Figure 8-2).

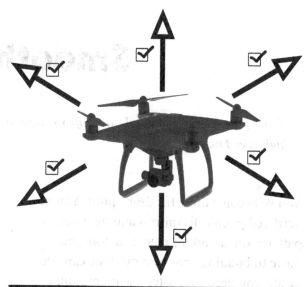

FIGURE 8-2 How a multirotor aircraft moves.

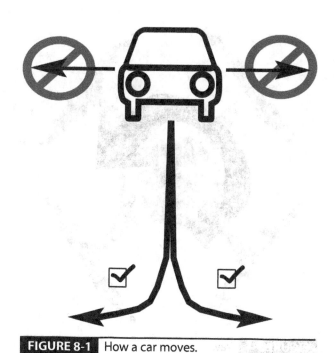

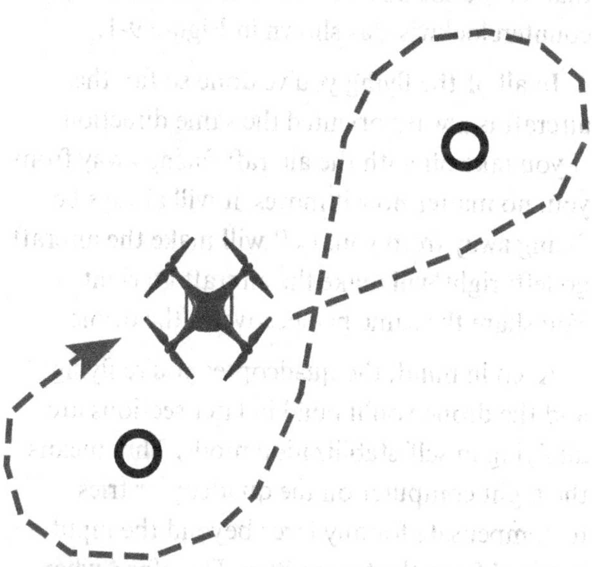

FIGURE 8-1 How a car moves.

FIGURE 8-3 Flying a figure-eight pattern.

Try to set up a figure-eight course. You can use two bedposts or two dining room chairs. Place them in a room three to eight feet apart. Set the drone in between the two obstacles, take off, and fly a figure-eight pattern as depicted in Figure 8-3.

Now, as you continue to fly, try changing the drone's altitude. Adding this new dimension can make flying more difficult at first. But remember: the drone is always facing the same direction. It might be moving up, down, right, left, forwards, and backwards, but it is always oriented the same way. In the next project, you'll become familiar with the last axes: yaw. You'll practice rotating the quadcopter as you fly it, increasing the maneuverability even more.

PROJECT 9

Smooth Sailing

You'll practice controlling the aircraft via yaw, allowing the drone to smoothly bank into turns and high-speed maneuvers.

You've become quite the drone pilot. You've mastered pitch/roll controls and figure-eight patterns on an indoor course. Before you move to building your own outdoor-capable drone, you need to master one more skill: yaw. To refresh your memory, yaw is the axis that allows the aircraft rotate clockwise and counterclockwise, as shown in Figure 9-1.

In all of the flying you've done so far, the aircraft is always oriented the same direction. If you take off with the aircraft facing away from you, no matter how it moves, it will always be facing away from you. Left will make the aircraft go left; right will make the aircraft go right. You share the same perspective as the drone.

Keep in mind, the quadcopter you're flying and the drone you'll build in later sections are all flying in self-stabilization mode. That means the flight computer on the quadcopter tries to compensate for any force beyond the input received from the transmitter. Therefore, when you introduce a new control axis like yaw, the flight computer will compensate for the pitched forward quadcopter, letting the drone rotate as you'd expect.

FIGURE 9-1 Yaw.

As mentioned, yaw lets you rotate the drone during flight. At first, yaw can be difficult to conceptualize and master. But as you practice more and more, it will become second nature. The easiest way to fly with yaw is to visualize yourself inside the quadcopter as opposed to an operator watching the drone from afar. You can always use the integrated LED lights to remind you of your orientation. On the Cheerson, blue is in front, and red is in back.

Since practice makes perfect, you should get started. Make sure your drone is fully charged. Set it on the landing pad, arm it, and get hovering.

Keep the right joystick centered (you can remove your thumb). To control yaw, you move the left joystick to the left and right. Be careful

not to move the left joystick up/down while you do that, as this motion will change the altitude! (Figures 9-2 and 9-3).

You'll see the drone rotate clockwise and counterclockwise. If you rotate the drone 180 degrees (half circle), and push the right joystick forward, the drone will fly in the opposite direction it was flying pre-rotation.

Try practicing flying with yaw. It may be frustrating at first, but once you get the hang of it, the possibilities are limitless. The drone will be significantly more agile: you'll be able to fly much more effectively.

Just like in the last project, try setting up a figure-eight course. Before you integrated yaw, the figure eight looked something like Figure 9-4.

FIGURE 9-2 Flying with yaw. Notice how the drone rotates.

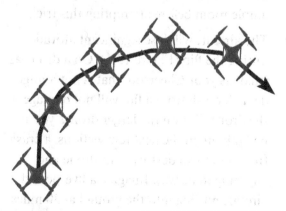

FIGURE 9-3 Flying without yaw. Notice how the drone does not rotate.

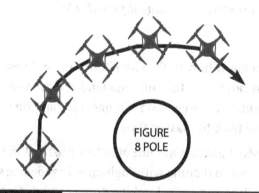

FIGURE 9-4 Figure 8 Pole without yaw.

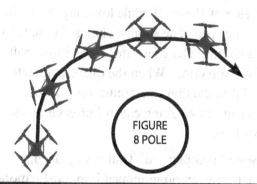

FIGURE 9-5 Figure 8 Pole with yaw.

With yaw, however, you'll be able to further master the figure-eight course. Flying should look similar to Figure 9-5.

You may be wondering why yaw matters. Why add the extra piece of complexity?

Once you start adding accessories to your drone, such as cameras, lights, and more, you must fly your drone in the orientation you desire. If you're capturing aerial footage, and the aircraft (and therefore camera) does not yaw/rotate, you'll miss your perfect video shot. Yaw keeps the front of the drone, and therefore camera, oriented in the direction of motion.

Now that you've mastered all of the axes, it's time to jump into something even more fun: flips! In the next project, you'll discover some of the drone's special capabilities and how they're able to perform cool tricks like flips.

PROJECT 10

Flips

Bored with normal flying? Make it more exciting. You'll practice sending the aircraft on wild adventures, including flips and rolls.

You know how to fly with pitch, roll, and yaw. You can hover, take off, and land. Before you graduate to larger, DIY drones, you have one more trick to master: flips!

Most aircraft are not able to withstand the stresses that come with high-speed maneuvers like flips and rolls. Luckily, drones are perfectly suited for these applications. When a flip occurs, the drone powers up two propellers on one side of the craft while lowering thrust from the other two. The sudden change in the thrust of each propeller sends the drone into a roll in a given direction. When the drone completes a full flip, the flight computer restabilizes the drone by retaking control of each of its propellers.

Sound complicated? That is why the flip maneuver is preprogrammed into most remotes for ready-to-fly drones. Before you actually complete a flip, you need to keep a couple of points in mind:

- The aircraft does not flip in a stationary place. It rapidly accelerates forward, backward, left, or right depending on the direction of the flip. Make sure you have ample room before attempting this trick.

- The aircraft will lose significant altitude during the flip (Figure 10-1). On a drone as small as your Cheerson, that is not a huge deal. A crash from a flip will not damage the drone. But on the larger drones you'll be building in the next few sections, a crash from a flip will destroy your drone and anything in its way. Imagine a five-pound aircraft hurling into the ground at 30 miles per hour!

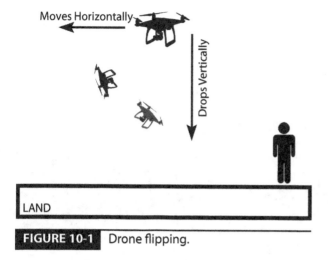

FIGURE 10-1 Drone flipping.

Flipping the Cheerson is easy. Every model is slightly different, so check the manual, but there will be a button you press to initiate "flip mode." Usually, you need to press down on the right joystick until you hear a beep or a button on the right side of the controller.

Charge your drone, set it on your landing pad, and arm it. Bring it to a hover one to two feet below the ceiling. Press the flip mode button, wait for the acknowledgment beep, and then move the right joystick in whichever direction you want the drone to flip. Release the right joystick, letting it move back to center. Watch as your drone completes a flip. Bear in mind the drone will lose altitude. The lower the battery level, the longer the drone will take to recover, and the more altitude it will lose.

Congratulations! You've now mastered indoor drone flight. You understand the basics of quadcopters, the different components, and the key safety issues to keep in mind. In the next sections, you'll work on building a much larger drone from scratch. This drone will be large enough to fly outdoors and carry accessories like cameras, lighting equipment, and others.

SECTION THREE
Picking the Parts

Now that you know how to fly, you can start building. Using the skills from the previous two sections, you'll put your new aerospace engineering skills to use and pick the perfect parts for your own high-performance drone.

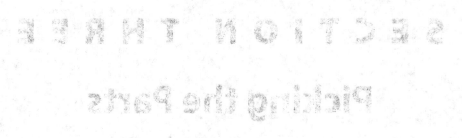

How Big?

When building your drone, choosing an appropriate size will be your most important decision. First, you'll explore multiple frame sizes.

Finally, the moment you've been waiting for is finally here: you can start building your first drone! Before starting any drone-build, you should think about the use case of the drone. Do you want a high-speed racer? A versatile drone for carrying around payloads? A cinematic rig? You're limited only by your imagination.

In this section, you'll build a midsized versatile quadcopter (four arms) capable of carrying a variety of payloads. The build process and techniques you use in the following sections apply to drones of all sizes. As you take an in-depth look at part selection in the next set of projects, you'll learn how to adjust for a larger or smaller drone. Overall, everything is pretty much the same—the biggest difference will be in frame size, propeller size, and battery size.

In general, frames can be made out of anything. They only need to serve a single purpose: keep all of the other drone components together and withstand the forces of flight! Drones have been made out of recycled clothes hangers, PVC pipe, and more. There's no wrong answer.

For the sake of simplicity, you can order a basic frame from a retailer like Amazon. To start, you should know the three most common frame sizes: 250, 330, and 450 mm. The 250 mm size is perfect for a lightweight racing drone. The 330 mm size is great for lifting a small payload. The 450 mm size is perfect for a versatile drone (Figure 11-1). These frames are well built, include mounting pads for the brushless motors, and have integrated

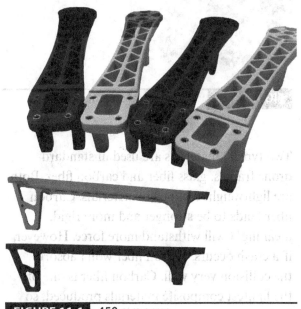

FIGURE 11-1 450 mm arms.

power distribution boards (we will discuss this component in depth when we assemble the drone).

The larger the frame, the longer each propeller can be; the propellers will end up generating more lift. More lift means your quadcopter is capable of flying with significantly heavier payloads. The downside, however, is drones that weigh more tend to be less agile and further subject to catastrophic wind conditions, which is discussed in depth later.

As you browse quadcopter frame options, you should note the following characteristics:

- Material
- Detachable Arms
- Foldable Arms

FIGURE 11-2   Motor mount on arm.

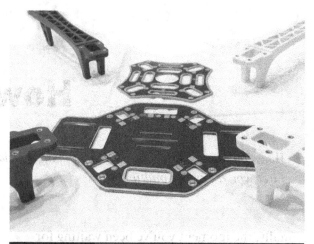

FIGURE 11-3   The other components.

Two typical materials are used in standard drone frames: glass fiber and carbon fiber. Both are lightweight composite materials. Carbon fiber tends to be stronger and more rigid, meaning it will withstand more force. However, if a crash occurs, carbon fiber won't absorb the collision very well. Carbon fiber is one of the lightest composite materials produced, so high-performance small drones may opt for this material. When you're just getting started, you really cannot go wrong with either material.

For your first frame, a glass fiber 450 mm build is recommended. A link to the frame used throughout the book is available at **diydronebook.com**.

The frame comes with several key components (Figures 11-2 and 11-3):

- Four (4) arms with integrated motor mounts
- One (1) top plate
- One (1) bottom plate with integrated power distribution board
- Multiple screws to connect the whole kit together

FIGURE 11-4   Exploded view of the 450 mm frame.

When your frame arrives, don't assemble it quite yet! You'll save that for after all of your parts have arrived. Since the parts are tightly integrated onto the drone, you must assemble components in the correct order (Figure 11-4).

Go ahead and order your frame. Your parts list should now coincide with Table 11-1.

Want to go even bigger on drone size? Well, you can. Drones can even fit a human! EHang, Inc., a Chinese-based drone company,

TABLE 11-1	Your Parts List
Part	Specification
Frame	450 mm glass fiber frame

REMINDER! Up-to-date parts are available at **diydronebook.com**.

recently unveiled a prototype of a drone that fits a human as shown in Figure 11-5. Once released, the drone will be able to fly one passenger for up to 20 minutes. The device has four "arms," each with two propellers, and should be able to hold up to 220 pounds. So is the EHang the start of a future where we all fly ourselves around in individual drones? Not exactly. The company says the drone will be 100 percent automated and essentially fly itself once users input a destination. While probably not as fun as operating it yourself, automation is definitely a safer alternative. While still many years away from coming to market, EHang's product is a fun glimpse of what the future of drone transportation could look like.

FIGURE 11-5 EHang's passenger drone.

In the next project, you'll be introduced to powertrain components and figure out the best set of electronic speed controllers (ESCs), brushless motors, and propellers.

PROJECT 12

Powertrain

With any aerospace project, the vehicle's powertrain is a critical aspect. How will it fly? How fast will it go? In this project, you learn the fundamentals of brushless motors, propellers, and ESCs.

Now that you've chosen a frame size for your drone, you can determine your powertrain components. First, you'll look at propeller options. Then you'll explore brushless motors and finally choose your ESCs.

The biggest constraint on powertrain is propeller size. If you're using a 250 mm frame, your propellers must be short enough so they

do not collide with one another when spinning. Propellers come in all shapes and sizes, but the most common propeller for quadcopters is a simple two-blade design. Occasionally you will see three-blade propellers, but those tend to be more expensive. The advantage of more blades is that the radius of the propeller is reduced for a given surface area, meaning you can generate

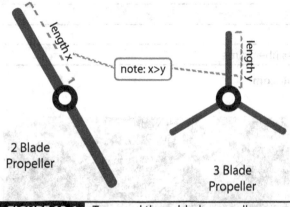

FIGURE 12-1 Two- and three-blade propellers.

more lift on a smaller frame. For your purposes, a simple (and less expensive!) two-blade propeller will work just fine (Figure 12-1).

Most propellers will have specifications regarding their compatible frame size. For example, a quick search for "quadcopter propeller" on Amazon brings up products that specify the frame size, as shown in Figure 12-2.

Assuming you are using the 450 mm frame that you'll be using throughout this book, Amazon has the perfect set of propellers. A link to the propellers is available at **diydronebook.com**.

In general, your propellers will break most often during flight. When your drone crashes, the propellers will bear the brunt of the impact; as a result, they are usually prone to snapping. Luckily, propellers are inexpensive and easy to replace. The set of propellers for the 450 mm frame comes with extras, just in case.

You'll notice propellers are labeled "clockwise" and "counterclockwise." On multirotor aircraft, half of the propellers rotate in the opposite direction of the others. This directionality is known as counter-rotation. Counter-rotation allows the quadcopter to automatically offset any tendency for yaw, based on the direction of the motors. In addition, by speeding up motors that spin in certain directions, the flight computer (FC) can make the drone yaw (rotate) on command (Figure 12-3).

The blades of the propellers are pitched in different directions as shown in Figure 12-4. This direction gives you a hint which way the propeller spins to generate lift.

Two propellers rotate must clockwise, and two must rotate counterclockwise. In this configuration, the propellers are counter-rotating one another. Remember from the first section that

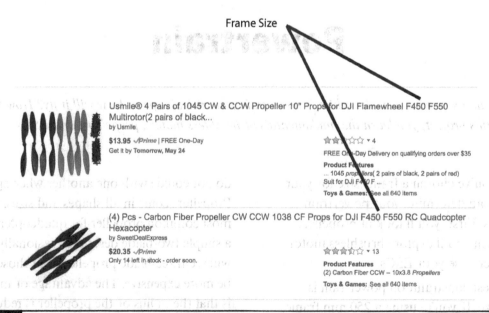

FIGURE 12-2 Amazon results for "quadcopter propeller."

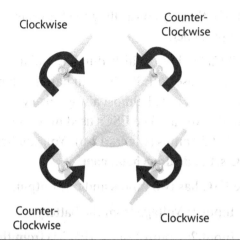

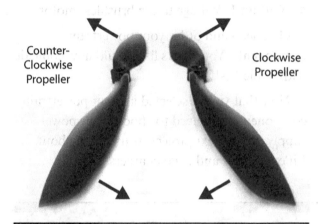

FIGURE 12-3   Direction of quadcopter propellers.

FIGURE 12-4   Clockwise versus counterclockwise propellers.

FIGURE 12-5   Brushless motor.

counter-rotation involves spinning a propeller or rotor in the direction opposite to that of another. Each propeller or rotor still produces all the same effects of torque, gyroscopic precession, slipstream, and P-factor. However, since all of these forces are dependent on the direction of a propeller or rotor's rotation, two rotors moving in opposite directions will produce these forces in opposing directions, causing most of them to cancel out. Therefore, when you power up your quadcopter, it will not rotate on its own until you direct it to yaw.

When you reach the assembly stage, you'll learn how to know which propeller to use on which arm. Now, you need to figure out which brushless motor to use.

Now that you have your propellers, you need to choose a brushless motor. Brushless motor power is determined by KV. KV is the number of rotations a motor will make under one volt of input. Remember, your drones run on LiPo batteries at around 11.1 volts. The motors used in this book are rated 935 KV, which means they spin at 10,378 rotations per minute under full load. That's really fast! Of course, the speed is why you need ESCs. The flight computer tells the ESCs exactly how fast to spin the propellers.

For quadcopters, brushless motors ranging between 600 and 3000 KV should work well on frame sizes of 250 to 450 mm. All brushless motors in this range have the same standardized mount, which makes it extremely easy to attach the motor directly to the frame discussed in the previous section (Figure 12-5).

To find the brushless motors used in this book, go to **diydronebook.com**.

Now that you have your propellers and brushless motors, you need your ESCs. ESCs regulate how much voltage reaches the brushless motor and control the rotations per minute

FIGURE 12-6   ESC example.

(RPM), thereby controlling the lift generated by each propeller.

ESCs tend to be fairly durable, and if you choose to use an ESC that can handle a larger quadcopter, it will almost always also work on a smaller drone. The ESCs used in this book are rated at 30 amps (Figure 12-6). You can find the ESC set at **diydronebook.com**.

The ESC has two inputs and one output:

- Input 1: Voltage from the battery
- Input 2: Control signal (PWM) from the flight computer
- Output 1: Voltage to the brushless motor

Go ahead and order your powertrain components. Your parts list should now coincide with Table 12-1.

Now that you've selected all your powertrain components, you need to choose your power supply. In the next project, you'll learn about LiPo batteries and LiPo chargers.

TABLE 12-1 Your Parts List	
Part	**Specification**
Frame	1x 450 mm glass fiber frame
Propellers	**4x + 8040 size propellers (half clockwise, half counter-clockwise)**
Brushless motors	**4x 935 KV brushless motors**
Electronic speed controllers (ESCs)	**4x 30 Amp ESCs**

REMINDER! Up-to-date parts are available at **diydronebook.com**.

PROJECT 13

Power

You can have a chassis and a beautiful powertrain setup, but how do you actually power the beast? Answer: with high-performance batteries, of course. You'll explore LiPo batteries, battery safety, and how to choose the perfect battery for the drone.

Without a power supply, your drone can't fly. As mentioned before, quadcopters are typically powered by LiPo batteries. These devices are similar to the batteries that power your laptop, phone, and more, except they tend to be packed in a very different shape. These batteries are extremely energy dense, and if treated improperly, they can be very dangerous.

Look at a typical LiPo battery, as depicted in Figure 13-1.

There are three key parts of the battery (Figure 13-2):

■ Main cable

■ Feedback cable

■ Capacity label

When you're ready to fly, you'll plug the main cable into your drone. The feedback cable remains unplugged. The main cable uses an XT60 connector.

When you're charging your battery, you must connect both the main cable and feedback cable to the charger. This setup allows the charger to monitor the battery status at all times. Using

the feedback cable, the charger is able to prevent overcharging (Figure 13-3).

The battery's capacity label describes two components: the number of cells and the milliampere hours (mAh). The number of cells represents how many smaller batteries make up the larger battery. For drones, you should use a 3S (three cell) battery, since it's the most common. The mAh is the most

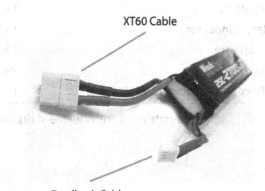

FIGURE 13-2 LiPo battery cables.

FIGURE 13-1 LiPo battery.

FIGURE 13-3 LiPo battery label.

TABLE 13-1	Battery Sizes
Frame Size	**Battery Specification**
250 mm	1300–2000 mAh, 3 cells
330 mm	2000–3500 mAh, 3 cells
450 mm	2500–4000 mAh, 3 cells

important metric. The larger the mAh, the heavier your battery but the longer the drone will stay airborne. We recommend the following batteries for the following frame sizes shown in Table 13-1.

For the 450 mm quadcopter you build in the next section, you'll use a 2700 mAh

battery. You can find a link to the battery at **diydronebook.com**.

When your battery arrives, make sure you store it in a cool, dry place. Leave the battery uncharged until you're ready to use it. Remember: safety first!

Go ahead and order your battery, charger, and assorted cables. Your parts list should now coincide with Table 13-2.

In the next project, you'll examine the remaining components: the transmitter and receiver. This vital link allows you to communicate with your drone instantaneously, commanding it to go on whatever mission you see fit.

TABLE 13-2	Your Parts List
Part	**Specification**
Frame	1× 450 mm glass fiber frame
Propellers	4× + 6040 size propellers (half clockwise, half counterclockwise)
Brushless motors	4× 935 KV brushless motors
Electronic speed controllers (ESCs)	4× 30 Amp ESCs
LiPo battery	**1× 11.1 V 3S 2700 mAh LiPo battery**
Battery connector	1× XT60 plug
Hookup wire	**5 feet 20 ga red/black hookup wire**
LiPo charger	1× LiPo battery charger
Flight computer power supply cable	**1× FC power supply cable**

REMINDER!   Up-to-date parts are available at **diydronebook.com**.

PROJECT 14
Radio Control

Every drone needs a way to listen to its operator! Now you'll choose a radio transmitter and receiver.

A drone's radio control component consists of two parts: the transmitter and receiver. The transmitter you'll use is very similar to the Cheerson CX-10 controller, except it's larger,

programmable, and has significantly more channels. In general, each drone you build can reuse the same transmitter/receiver pair. It's very easy to swap out the components for each

FIGURE 14-1 6 Channel controller.

Trim Switches

FIGURE 14-2 Trim switches.

drone, which saves you from paying for these additional parts every time you build another quadcopter.

The transmitter used in this build is a six-channel controller. As shown in Figure 14-1, it contains a few more buttons and switches than the Cheerson controller.

The joysticks and on/off button should look familiar. You'll notice a lot more buttons: trim switches for each joystick, rotating knobs, quick flip switches, and more. The trim switches are shown in Figure 14-2. They function exactly as they did on the Cheerson controller.

The additional knobs and switches are entirely programmable. This controller only has six channels available, and four of them are used by the joysticks. But the other two channels are completely within your control. In the next section, you'll learn how to program them and connect them to two of the unused knobs and buttons.

The last component that comes with the controller is the receiver. The receiver, shown in

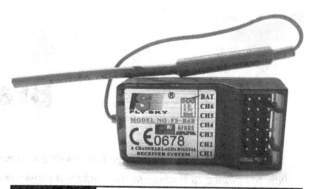

FIGURE 14-3 Receiver.

Figure 14-3, will mount to the frame of the drone. It will connect directly to the flight computer, relaying the controller's inputs and commands.

Go ahead and order your radio set. Your parts list should now coincide with Table 14-1.

In the next project, you'll familiarize yourself with the final piece of equipment: the flight computer. The flight computer serves as the quadcopter's brain and connects all the components together.

TABLE 14-1 Your Parts List

Part	Specification
Frame	1× 450 mm glass fiber frame
Propellers	4× + 6040 size propellers (half clockwise, half counterclockwise)
Brushless motors	4× 935 KV brushless motors
Electronic speed controllers (ESCs)	4× 30 Amp ESCs
LiPo battery	1× 11.1 V 3C 2700 mAh LiPo battery
Battery connector	1× XT60 plug
Hookup wire	5 feet 20 ga red/black hookup wire
LiPo charger	1× LiPo battery charger
Flight computer power supply cable	1× FC power supply cable
Radio set	**1× 6-channel digital receiver and transmitter**

REMINDER! Up-to-date parts are available at **diydronebook.com**.

P R O J E C T   1 5

Brains

You're almost there. You just need one more component: the brains! The drone needs to know what do to do, how to balance, and where to fly. Luckily, a single component handles all of these needs: the flight computer.

The flight computer is the single-most important part of any drone. As shown in Table 15-1, a typical flight computer has a variety of inputs and outputs.

A lot of different flight computers do exist, and, at the end of the day, they all serve the same purpose. In fact, some drone fliers even opt to build their own FCs. However, with high-end FCs costing less than $20, using an off-the-shelf solution seems most logical.

In the next section, you'll be using a CC3D flight computer. The CC3D is configured from a Windows or Mac computer via a standard

TABLE 15-1 Flight Computer (FC) Inputs and Outputs

Input	Output
Radio receiver	ESCs (and therefore brushless motors)
Battery voltage	Camera switches and gimbals (optional)
Gyroscope (integrated in FC)	Landing gear (optional)
Accelerometer (integrated in FC)	Other accessories (optional)
USB port	Laptop (during configuration only)

FIGURE 15-1 CC3D flight computer.

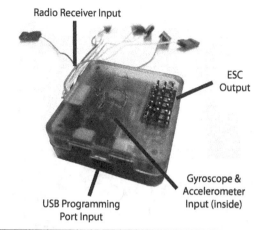

Radio Receiver Input

ESC Output

Gyroscope & Accelerometer Input (inside)

USB Programming Port Input

FIGURE 15-2 CC3D inputs and outputs.

micro or mini USB cable, which makes it much easier to set up than having to use a tiny screen integrated directly onto the FC. The CC3D flight computer is shown in Figure 15-1.

The flight computer comes with a variety of cables, most of which connect the flight computer to the radio receiver (Figure 15-2). Don't worry: you'll learn how to set up, program, and configure the flight computer in the next section.

For now, go ahead and order a flight computer. To buy the CC3D model that you'll use in the next section, go to **diydronebook.com**. Your parts list should now coincide with Table 15-2.

In the next section, you'll take all the parts you've ordered and bring your drone to life!

TABLE 15-2 Your Parts List	
Part	**Specification**
Frame	1× 450 mm glass fiber frame
Propellers	4× + 6040 size propellers (half clockwise, half counterclockwise)
Brushless motors	4× 935 KV brushless motors
Electronic speed controllers (ESCs)	4× 30 Amp ESCs
LiPo battery	1× 11.1 V 3C 2700 mAh LiPo battery
Battery connector	1× XT60 plug
Hookup wire	5 feet 20 ga red/black hookup wire
LiPo charger	1× LiPo battery charger
Flight computer power supply cable	1× FC power supply cable
Radio set	1× 6-channel digital receiver and transmitter
Flight computer	**1× CC3D flight computer**

REMINDER! Up-to-date parts are available at **diydronebook.com**.

SECTION FOUR
Assembly

Once your components arrive, you can start putting together your drone!

Soldering 101

The only specialized tool you'll require is a soldering iron. Soldering may seem intimidating, but with this easy-to-understand guide, you'll be ready to go in no time! You'll practice by creating a basic flashlight.

Before you dive into assembling the drone, you still need to master one more skill: soldering. Soldering involves melting a conductive material, called solder, to bond together two electrical connections. Soldering is used in electrical circuits and can occur in many different shapes and sizes: from computer-controlled soldering ovens that connect thousands of components at once to human-controlled soldering irons.

In this project, you'll learn how to use a manual soldering iron. You'll assemble a basic LED flashlight. If you already know how to solder, you can skip this project.

You will need four components for this project, as listed in Table 16-1.

Order all of your components. Once they arrive, lay everything out as depicted in Figure 16-1.

The flashlight is extremely simple and will teach you the basics of soldering. First, look at the battery connector. You'll notice two wires that extend from it: one black and one red as shown in Figure 16-2. In electronics, red wires

always represent positive, and black wires always represent negative (also known as ground). This book won't teach you electrical engineering, but at a high level the positive and negative wiring represents the direction electricity flows from the battery into the LED.

Next, you need to identify the positive and negative leads on the LED. If you look at the LED carefully, you'll notice one side is slightly flattened. That flattened side is the negative half, and the lead closest to that side is your negative lead (Figure 16-3).

Now that you know the positive and negative wires on both the LED and battery holder, you need to connect them. You never want to accidentally solder-touch both the negative and positive side at the same time when a circuit is powered, a problem known as an "electrical short." To prevent that, bend the leads on the LED so they are further apart, as shown in Figure 16-4.

Take the battery connector and look closely at the two wires. You'll notice the red and

TABLE 16-1	Components Required	
Item	**Description**	**Usage**
Soldering iron kit	Basic soldering iron and solder	Drone assembly and flashlight project
LED + resistors	Red LED with resistors	Flashlight project *only*
Battery holder	AA 2× battery holder with leads and integrated switch	Flashlight project *only*
Batteries	2× AA batteries	Flashlight project *only*

REMINDER! All of these components are available for order via Amazon on **diydronebook.com**.

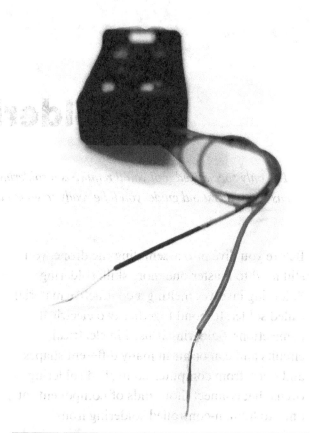

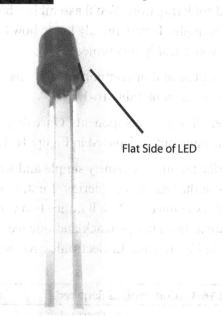

FIGURE 16-2   Positive and negative wires.

Flat Side of LED

FIGURE 16-1   Flashlight components.

FIGURE 16-3   LED polarity.

black coloring is actually a rubber coating on top of the inner metal wire. The rubber acts as an insulator, preventing an electrical short. However, you need to expose more metal wire in order to connect the battery connector to the LED. To do that, take a pair of scissors (or your fingernails!) and gently scrape on the plastic coating until you expose more metal wire.

The resistor is the other component that came with your LEDs. The resistor prevents too

much electricity from going to the LED. Unlike the battery pack and LED, the resistor can be installed in either direction, meaning it lacks polarity.

FIGURE 16-4 Bent LED leads with resistor.

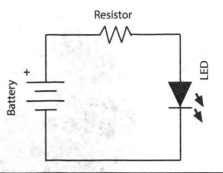

FIGURE 16-6 Flashlight schematic.

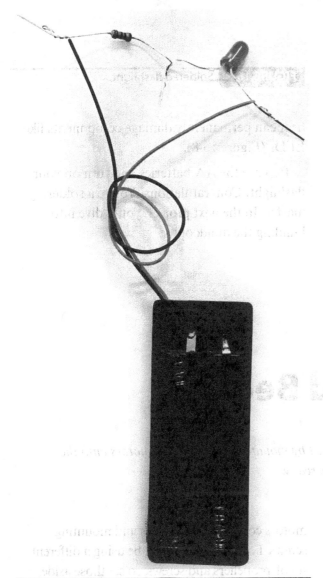

FIGURE 16-5 Completed circuit, pre-solder.

Wrap the negative wire from the battery connector around the negative lead from the LED. Wrap the positive lead from the LED around one end of the resistor. Do the same with the positive wires, as depicted in Figure 16-5.

Test your circuit! Put batteries in the battery holder and flip the switch. Does the LED turn on? If so, congratulations. Your circuit works. If not, try flipping the connections and make sure the negative LED lead is connected to the negative wire, and the positive LED lead is connected to the positive wire. Figure 16-6 depicts your circuit in an electrical schematic.

Remove the battery from your flashlight and plug in the soldering iron. You're about to master a new skill. Your soldering iron will get extremely hot, so only touch it by the handle. The iron will take about five minutes to become hot. While its heating, make sure the metal portion doesn't touch anything heat sensitive like a wooden table or any fabric material. Most soldering irons come with stands—if yours did, prop it on the stand while the iron heats.

In the meantime, get your solder ready. Pull out the tip of the solder and set it up as shown in Figure 16-7.

Once your soldering iron is ready, pick up the iron with your right hand and solder with your left hand. Touch the tip of the iron, the end of the solder, and the positive LED lead/battery wire connection together at the same time. You'll see the solder melt and wick around the lead and wire, bonding the connection.

Remove the soldering iron and let the connection cool for 30 seconds. The solder will harden, and your first soldering job is complete. Repeat the same technique with the negative lead

FIGURE 16-7   Solder ready for soldering.

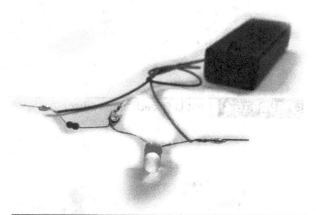

FIGURE 16-8   Soldered flashlight.

and wire. If you make a mistake, don't worry: solder can be remelted with the iron.

Try to minimize the amount of time the iron spends touching the electrical components. The iron is very hot, and prolonged exposure to that heat can permanently damage components like LEDs (Figure 16-8).

Replace the AA batteries and turn on your flashlight. Congratulations! You're a soldering master. In the next project, you'll dive into building the quadcopter.

PROJECT 17

The Build Begins!

Now you'll begin to assemble your drone. You'll start by mounting the brushless motors onto the frame and getting the electronic speed control units ready.

For this project, you'll need the following parts:

- 4× quadcopter frame arm
- 16× quadcopter frame mounting screws
- 4× brushless motor
- 4× electronic speed controller

Open the frame kit, and you'll notice eight sets of parts: four arms, a power distribution board, a frame body, and two sets of screws. Next, open the brushless motor sets you ordered. Some brushless motors come with propellers and mounting screws. For this build, you'll be using a different set of propellers and screws, so set those aside.

First, look at the brushless motor. At the top of the motor's pole, you'll notice a nut that's used to secure the propeller onto the motor, as depicted in Figure 17-1. You'll notice the nuts on the top of the propeller come in two sets of colors. If you remember from the previous section, two of the drone's propellers

FIGURE 17-1 Brushless motor and propeller nut.

FIGURE 17-3 Mounting the motors onto the arms.

FIGURE 17-2 Screws from the frame set.

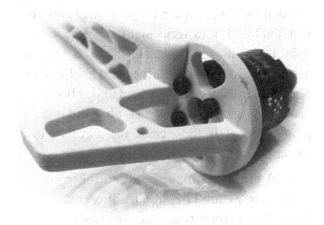

FIGURE 17-4 Tighten the motors.

spin clockwise, and two spin counterclockwise. These different-colored nuts will show you which motors spin in which direction. If you take a close look at the nuts and the motor pole, the threads are oriented in different directions for clockwise and counterclockwise. Later in this section, you'll understand the direction in which each motor needs to spin.

Turn the motor over, and you'll notice four mounting screw holes. The screws that came with your quadcopter frame will fit these holes perfectly. Identify the correct set of screws, as shown in Figure 17-2.

Mount the brushless motors onto the four arms, as depicted in Figure 17-3.

To make life easier in the future, attach the motors with red propeller nuts to the red arms. Attach the motors with black propeller nuts to the white arms. Following

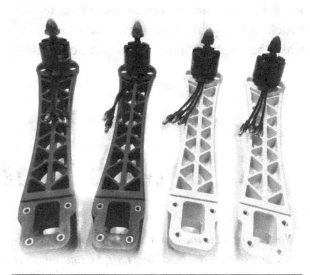

FIGURE 17-5 Four arms and four brushless motors.

this technique will help insure your motors are spinning in the right direction as you assemble the drone throughout the section (Figures 17-4 and 17-5).

FIGURE 17-6   Connecting the ESC.

Now that the motors are mounted onto the arms, stick the wires coming from the motor through the frame. When the propeller spins up, you don't want it to accidentally sever the wires. Take a close look at the end of the wires—you'll notice they seem to be attached to a gold-colored plug. The brushless motors used in this book come pre-attached to the quick-plug, but other motors do not. In those cases, you may have to solder the plugs yourself.

Look at the electronic speed controller, as depicted in Figure 17-6. You'll notice that the ESC has inputs that match the three plugs coming from the brushless motor. Try temporarily plugging them in now. The order in which the wires are connected doesn't matter. By switching two of the three wires, you'll be able to make the brushless motors spin in the opposite

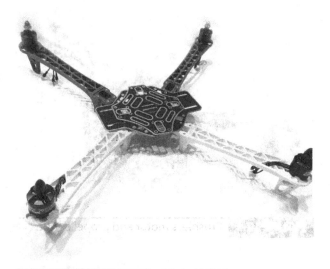

FIGURE 17-7   Work in progress.

direction. Again, this doesn't matter for now—you'll be able to adjust this later in the section.

Unplug the ESC from the brushless motor, and you'll see your quadcopter start to come to life. You can get a feel for the size of the craft. Your drone should resemble Figure 17-7.

Now that you know how the parts fit, you'll solder the ESCs to the power distribution board in the next project. You'll also connect the battery plug to the power distribution board and mount the board to the frame.

PROJECT 18

Now *That's* a Drone

In this project, the drone starts to come alive. You'll assemble the frame and mount the electronic speed controllers, brushless motors, and power distribution board (Figure 18-1).

You'll be using the parts shown in Table 18-1.

Review the power distribution board. You'll notice five sets of positive and negative connectors. Four of those sets are for the

ESCs, and the remaining set is for the battery plug.

Solder the XT60 battery plug to the positive/negative set depicted in Figure 18-2. Remember:

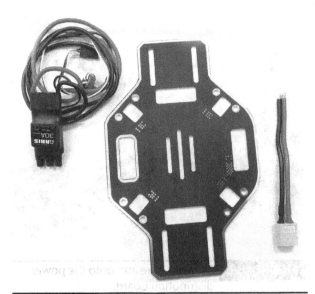

FIGURE 18-1 Power distribution board, battery plug, and ESC.

Connections

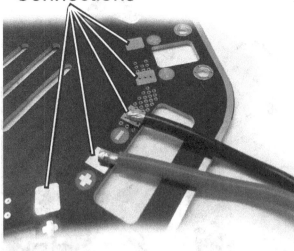

FIGURE 18-2 Power distribution board connections.

TABLE 18-1   Parts Used in This Project

Quantity	Part
4	Arm + brushless motor
1	Power distribution board
1	XT60 battery plug
4	Electronic speed controller

red is positive, and black is negative. When completed, your power distribution board should resemble Figure 18-3.

Next, you'll connect the ESCs. Your ESCs may have very long red and black cables; if that's the case, feel free to use wire cutters to trim the cables down to six inches in length. You'll then need to strip a centimeter off of the end of each wire, as explained in the previous project. Once your ESCs are ready, solder them as depicted in Figures 18-4 and 18-5.

Now that all of your ESCs are connected to the power distribution board, screw the power distribution board to the four arms. The screws for connecting the frame came with your frame

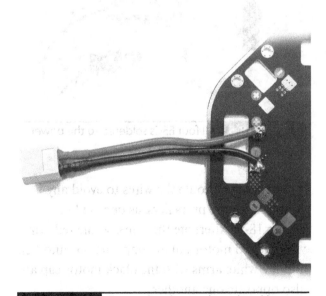

FIGURE 18-3 Power distribution board and battery connector.

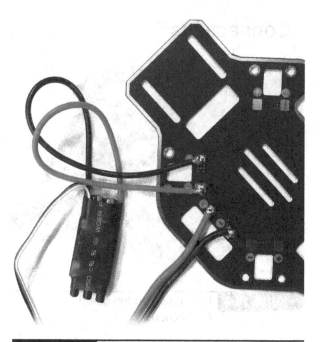

FIGURE 18-6 Screwing one arm onto the power distribution board.

FIGURE 18-4 One ESC soldered to the power distribution board.

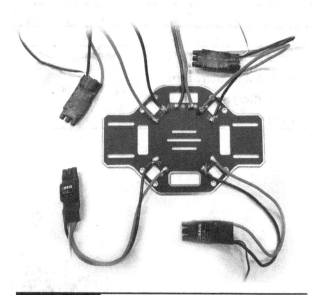

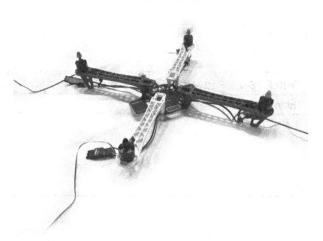

FIGURE 18-7 Four arms connected.

FIGURE 18-5 All four ESCs soldered to the power distribution board.

kit. Be sure you route the wires to avoid any contact with the propellers, as depicted in Figure 18-6. Alternate the arms, so the red arms with the red motor nut are opposite one another, and the white arms with the black motor cap are also opposite one another.

Once you have attached all four arms to the power distribution board, your quadcopter should resemble the drone in Figure 18-7.

Plug your ESCs into your brushless motors. Make sure the wires have some slack but not enough to hang low or get caught on any part of the drone. Secure the ESC to the frame using zip ties or tape.

In the next project, you'll attach the flight connector to the drone's frame.

PROJECT 19

Flight Computer

You'll mount the drone's brain, the flight computer, on your drone.

In this project, you'll use the parts shown in Table 19-1.

Before you mount the flight computer on the drone, you need to figure out which side of the drone is the front. To do that, look at one of the arms. Examine the motor from above and turn the propeller nut on top of the motor shaft. Turn the nut clockwise. If the nut tightens onto the shaft, label the arm with a piece of tape as **counterclockwise (CCW).** Otherwise, mark it **clockwise (CW).** This process is depicted in Figure 19-1.

Repeat the above process for each of the four arms. When you're done, be sure each arm is labeled, similar to Figure 19-2.

Next, refer to the diagram in Figure 19-3. Using the direction label on each arm, figure out which side of the drone is the front. Label this side as "front," so you don't forget.

Take the CC3D out of its box and attach the receiver cable, as shown in Figure 19-4. You'll attach this cable to the receiver, which you'll install in the next project. As you look at the CC3D, notice a small arrow on the top of the unit. The arrow points in the direction of the front of the drone.

Rotate Clockwise

Nut Tightens? Label **CCW**
Nut Loosens? Label **CW**

FIGURE 19-1   Clockwise versus counterclockwise.

FIGURE 19-2   Labeled arms.

It's important to mount the CC3D facing the right direction. The sensors within the unit all calibrate with directionality in mind. Before you mount the CC3D, you need to attach the frame's top plate to the drone.

TABLE 19-1	Parts Used in This Project
Quantity	**Part**
1	CC3D flight computer
1	Frame kit: top frame

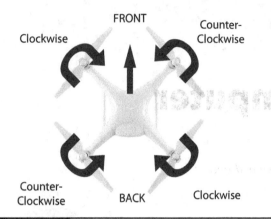

FIGURE 19-3 Figure out which side is the front.

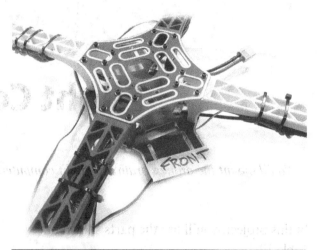

FIGURE 19-5 Top frame.

FIGURE 19-4 CC3D and receiver cable.

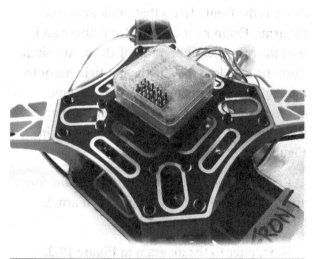

FIGURE 19-6 CC3D mounted.

Take the top plate of the frame, as shown in Figure 19-5, and attach it to the top of the arms.

To attach the CC3D to the frame, use double-sided sticky tape. Since the unit's orientation needs to remain constant for self-stabilization to work, you want to make sure the flight computer is firmly mounted. Mount the flight computer as shown in Figure 19-6 on the center of the top frame with the arrow on top of the CC3D facing the front of the quadcopter. You may opt to mount the CC3D onto a padded surface, like a piece of foam.

You don't yet have to plug in any wires or cables—you'll do that in the next project. The drone is almost complete.

PROJECT 20

Looks Matter

Last but not least, you'll mount the battery onto the drone and set up a cable management system to ensure the drone's high-speed propellers don't damage it.

In this project, you'll use the parts shown in Table 20-1.

One key piece remains: the radio receiver. Using a piece of two-sided tape, set your receiver on the front end of the drone, as shown in Figure 20-1.

The receiver has an antenna, which can be secured next to the receiver itself using zip ties, as shown in Figure 20-2.

Now that all your components are in place, you need to start wiring everything. Start with the receiver to flight computer cable. Take the input cable that comes with the CC3D and plug it into the receiver. One of the cables will be red, black, and white; this one is the Channel 1 input. From there, the wire next to the red/black/white cable should be Channel 2, and the one that follows will be Channel 3, then Channel 4, and so on. Make sure you plug the cables in correctly, as shown in Figure 20-3.

Once your cables are plugged into the receiver, snake the other end through the frame to the CC3D and plug it in.

Finally, you need to attach the ESC control cables. The motors should be laid out as shown in Figure 20-4.

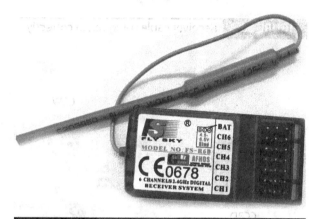

FIGURE 20-1 Receiver.

FIGURE 20-2 Secured with zip ties.

TABLE 20-1	Parts Used in This Project
Quantity	**Part**
1	CC3D flight computer
1	Frame kit: top frame
1	Flight computer power supply cable

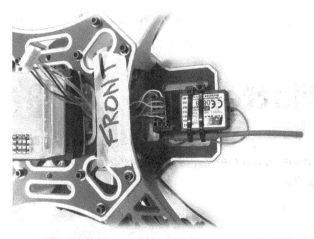

FIGURE 20-3   Receiver cables plugged in correctly.

FIGURE 20-5   CC3D motor outputs.

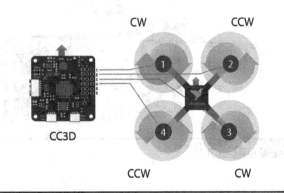

FIGURE 20-4   Motor layout.

You should plug the first motor into Channel 1 on the CC3D, the second motor into Channel 2, and so on. Figure 20-5 shows the plugs. Again, make sure they are in the correct orientation.

The CC3D flight computer still needs a way to receive power from the battery. The flight computer power supply cable plugs in between the battery and the quadcopter, converting some of the energy to five volts, which is the voltage that powers the CC3D. Take the output cable from the power supply cable and plug it into either of the open ports on the CC3D, maintaining the same orientation as the other cables.

Now you'll tidy everything. Using zip ties, secure any loose wires to the frame—you don't want the propellers to sever them.

You just connected many parts. Before you configure the flight computer, take a moment to understand what happens when the drone flies.

The battery sends power, through the flight computer power supply cable, to the CC3D. The battery also sends power to the ESCs, but they stop any electricity from reaching the motors and prevent them from spinning. The CC3D, through its input wire, sends power to the receiver.

The flight computer, acting as the brain, measures the aircraft's position and tilt using its integrated accelerometer and gyroscope. In combination with input from the receiver, the flight computer tells each ESC to release some voltage to the brushless motor, spinning the attached propeller to generate more lift.

When everything runs in perfect synchrony at high speeds, your drone is able to fly.

In the next section, you'll configure your drone and get it flying!

SECTION FIVE
We Have Liftoff

In this section, you'll program the drone, calibrate the motors, and start flying!

PROJECT 21

It's Alive!

Using an open-source, ground-control, computer-based tool, you'll plug the drone into a Windows or Mac computer and set up the flight computer.

Before you program the flight computer, make sure that your drone's components are hooked up correctly. Refer to Figure 21-1 for the overall setup.

First, confirm the electronic speed controllers' PWM cables are plugged into slots one through four on the CC3D flight computer. Make sure they're all the correct orientation: black closest to the edge, red in the middle, and white closest to the center of the flight computer, as shown in Figure 21-2.

Second, make sure the CC3D has power. Your flight computer power supply cable should have one end plugged into the XT60 battery plug you soldered, one end not plugged into anything (the battery will eventually go here),

and the PWM cable plugged into slots five or six of the CC3D.

If your power supply cable doesn't come with a PWM cable, don't worry—that's an easy fix! You'll need to order an extra set of PWM cables, which you can find at **diydronebook.com**. Once the cable arrives, cut off the PWM connector and leave about six inches of cable attached. You'll find three wires: red, white, and black. Ignore the white wire (Figure 21-3).

Now, disconnect the existing removable cable from the power supply. Near that port, on the power supply you should see four pinholes. One will be labeled "5V" or "+." Another will be

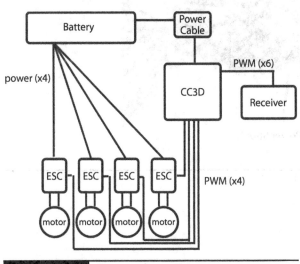

FIGURE 21-1 Drone schematics.

FIGURE 21-2 ESCs connected to the flight computer.

Black: – or **gnd**

Red: + or **5 V**

White: ignore

FIGURE 21-3 Modify PWM cable.

labeled " −" or "gnd" as shown in Figure 21-4. Take the PWM cable, stick the red wire through the "5V" or "+" hole and solder it in place. Take the black wire, stick it through the "−" or "gnd" hole and solder it in place.

Finally, make sure the input cable on the CC3D is connected properly to the receiver. There should be six plugs coming from the cable. The plug with three wires goes into Channel 1, the next plug on the cable goes to Channel 2, and so on.

Now that everything is correctly assembled, charge your LiPo battery according to the instructions on the charger you purchased. While your LiPo charges, install LibrePilot, the free open source Unmanned Aerial Vehicle configuration tool. A link to download LibrePilot is available at **diydronebook.com**. LibrePilot is available for Windows, Mac, and Linux operating systems.

In the next project, you'll connect the CC3D to the computer and calibrate the drone.

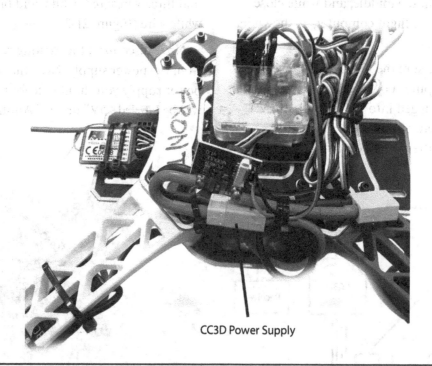

CC3D Power Supply

FIGURE 21-4 Power supply and CC3D.

PROJECT 22

Calibration Is Key

Using a computer, you'll calibrate the electronic speed controllers, brushless motor idle speeds, gyroscope sensitivity, and accelerometer angle.

Now that everything is hooked up correctly, you'll calibrate the drone. First, make sure you've removed your drone's propellers and battery. Next, plug in a USB-Mini cable into the CC3D and computer. USB Mini cables are available at **diydronebook.com**.

Open LibrePilot. You'll see a screen similar to Figure 22-1.

Click "Vehicle Setup Wizard" to begin. You'll be reminded to remove the propellers

and disconnect the battery (which you should have already done!) as shown in Figure 22-2.

Next, LibrePilot will update the CC3D firmware to the newest available version. Go through the prompts, as shown in Figure 22-3, to update your flight computer.

LibrePilot will identify your CC3D model, as shown in Figure 22-4.

FIGURE 22-1   LibrePilot home screen.

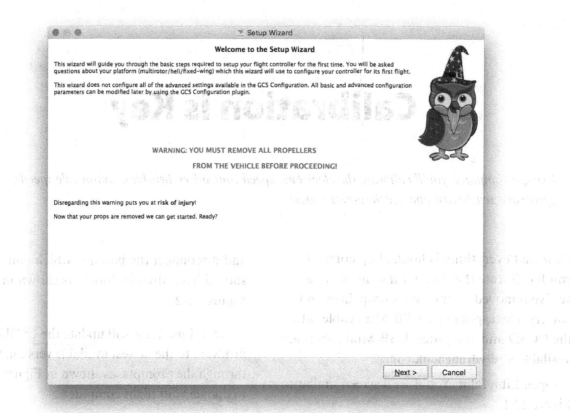

FIGURE 22-2 Vehicle setup.

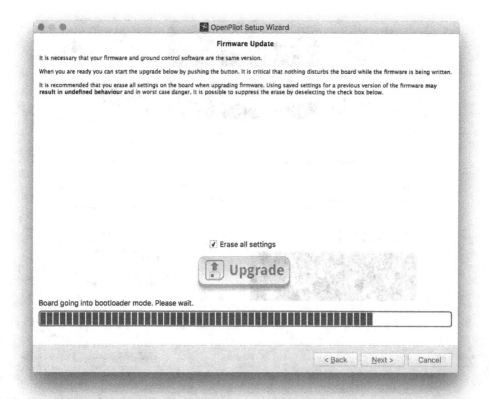

FIGURE 22-3 Updating firmware.

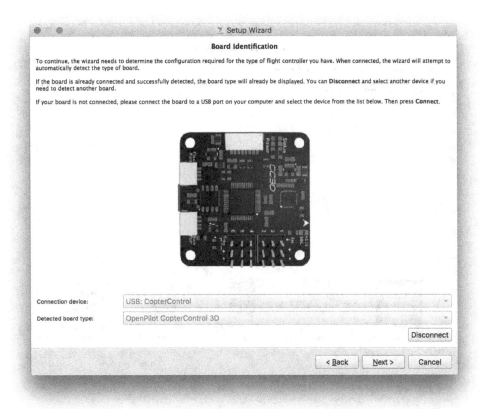

FIGURE 22-4 CC3D identification.

You'll then be prompted to specify details about your drone. Select "PWM" and "Multirotor" (Figures 22-5 and 22-6).

Next, you'll be prompted to choose your multirotor configuration. You'll be flying in "Quadcopter X." As you quickly become an expert on multirotor aircraft, you'll be able to explore the other drone options (Figure 22-7).

LibrePilot will now ask you for your ESC type. The ESCs used throughout this book are high-performance Rapid ESCs (Figure 22-8).

Now, you need to set your drone on a level and stable surface. LibrePilot needs to calibrate the gyroscope and accelerometer, as shown in Figure 22-9.

The gyroscope and accelerometer are now calibrated! Next, LibrePilot will need to calibrate

the ESCs to ensure that all propellers spin at the same rate on the CC3D's command. Follow the steps shown in Figure 22-10. You'll need to connect and disconnect your LiPo battery in this step.

Finally, it's time to spin up those motors! You'll need to reconnect your LiPo battery to the drone. In this step, you'll figure out the idle rate of the motors. The idle rate is the maximum amount of energy that can be applied to the motor before it begins to spin. You'll also double check that each motor is spinning in the correct direction. Follow LibrePilot's prompts, as shown in Figure 22-11.

Spinning the Wrong Way?

If any of the four motors spin in the wrong direction, don't worry! You simply need to reconnect the motor to the ESC, swapping two of

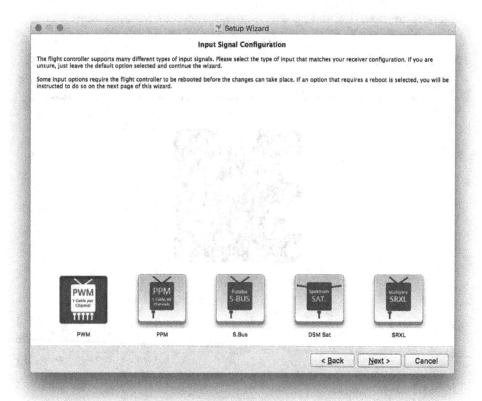

FIGURE 22-5   PWM connections.

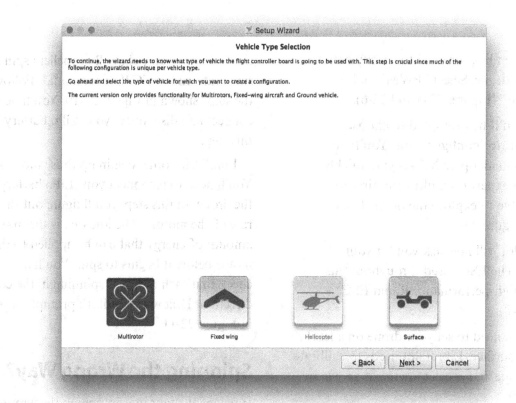

FIGURE 22-6   Multirotor.

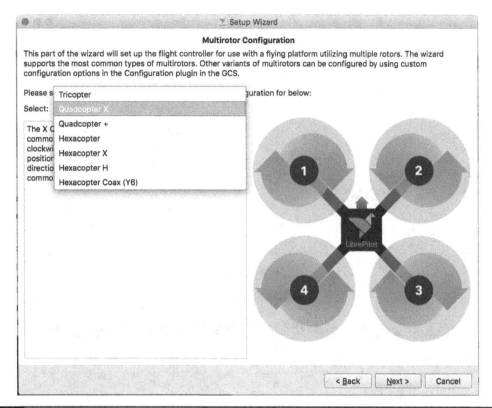

FIGURE 22-7   Quadcopter X.

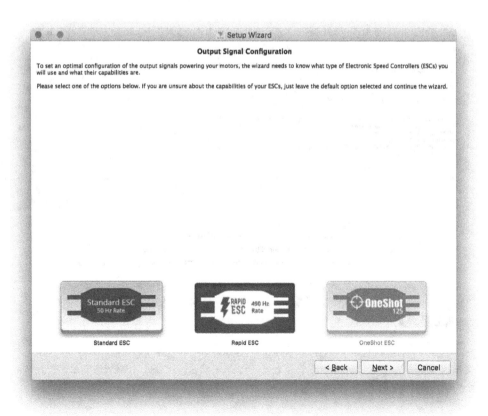

FIGURE 22-8   ESC type.

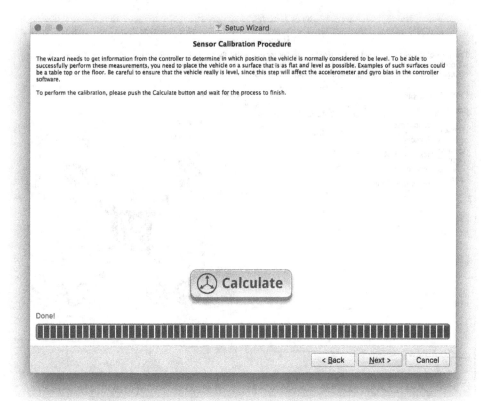

FIGURE 22-9   Sensor calibration.

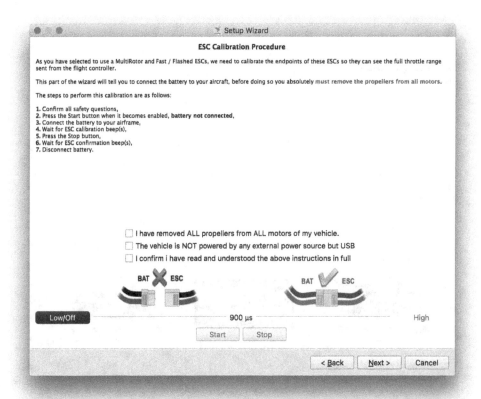

FIGURE 22-10   ESC calibration.

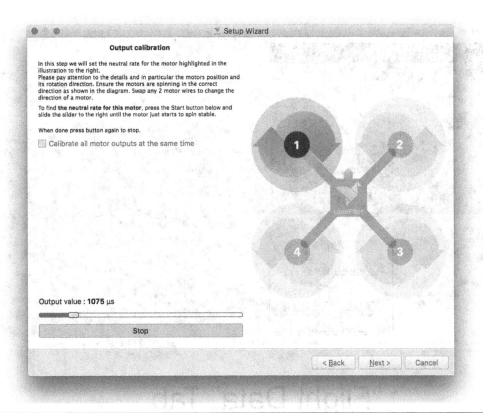

FIGURE 22-11   Motor spinning.

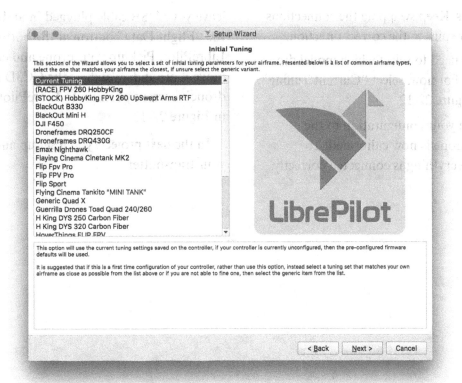

FIGURE 22-12   Current tuning.

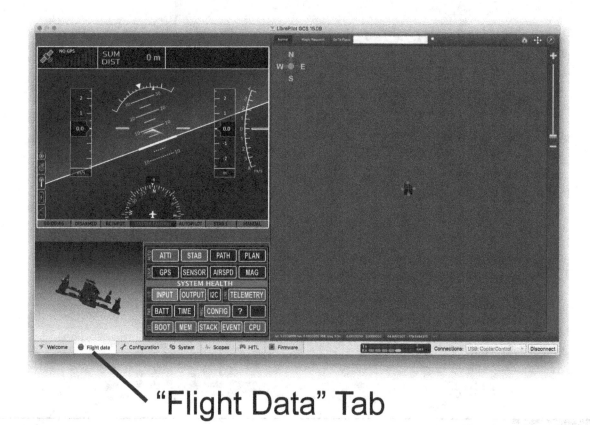

"Flight Data" Tab

FIGURE 22-13 Successful setup.

the connections. Keep swapping the connections until the motor spins in the correct direction.

Finally, you need to select which model drone you are flying. For now, select "Current Tuning" as shown in Figure 22-12.

Finally, save your configuration to the CC3D. Your drone is now calibrated! To confirm that everything is connected correctly,

leave your USB cable plugged in and click the "Flight Data" button at the bottom of LibrePilot. Pick up your drone and carefully move it around. Watch as the CC3D relays the drone's current position to LibrePilot, as shown in Figure 22-13.

In the next project, you'll set up and calibrate your transmitter.

Get in Control

You'll calibrate and connect your transmitter using the free and open-source software, LibrePilot.

Now that your CC3D is calibrated and turned on, you need to also calibrate your transmitter. To start, make sure your CC3D is plugged into your computer and that the drone is powered on with its LiPo battery attached. Open LibrePilot, go to "Configuration," and start the "Transmitter Setup Wizard" as shown in Figure 23-1.

Proceed through the first few steps and leave the default options checked as shown in the following figures. You'll notice you confirm

you're flying in Mode 2, which is the way you learned to fly the Cheerson CX-10 in Section One (Figures 23-2 and 23-3).

Power on your transmitter once your reach the screen shown in Figure 23-4.

LibrePilot asks you to move the Throttle stick up and down. Don't move any other controls. Once LibrePilot recognizes the Throttle, it will ask you to do the same for the roll control, as shown in Figure 23-5.

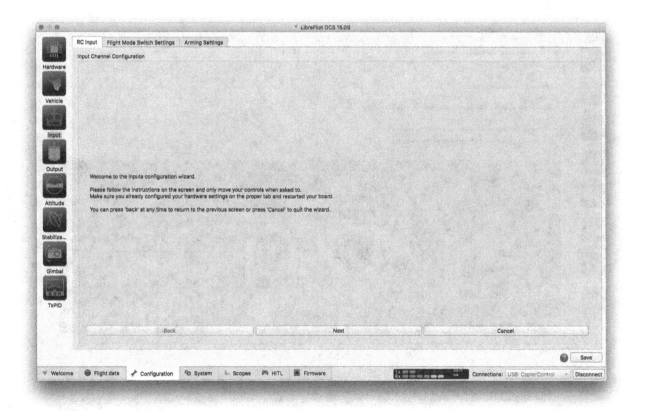

FIGURE 23-1   Transmitter setup wizard.

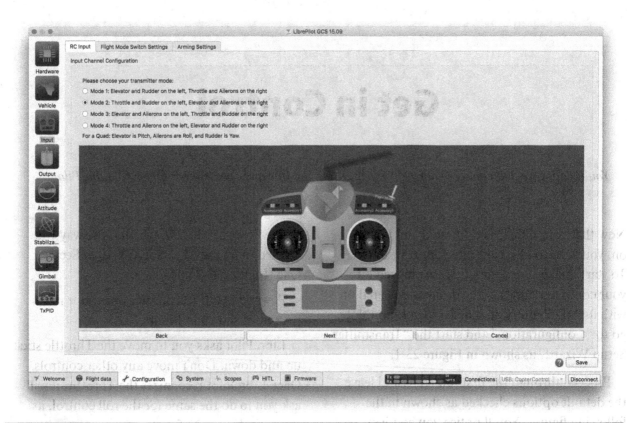

FIGURE 23-2 Mode 2.

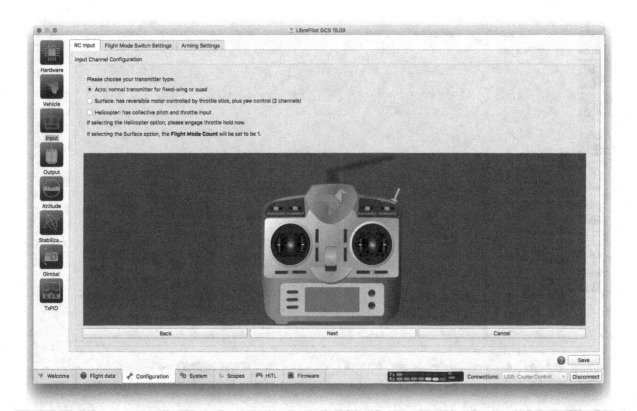

FIGURE 23-3 Acro mode.

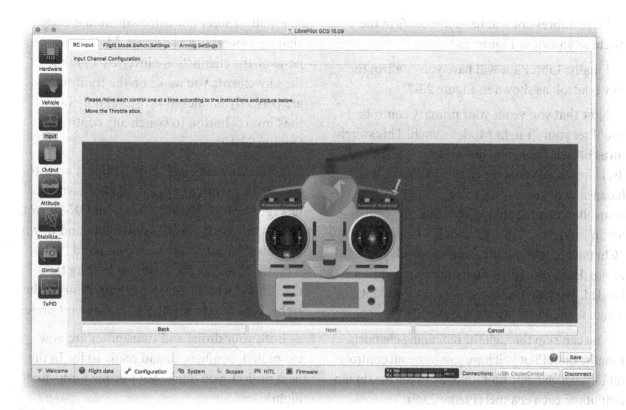

FIGURE 23-4   Throttle check.

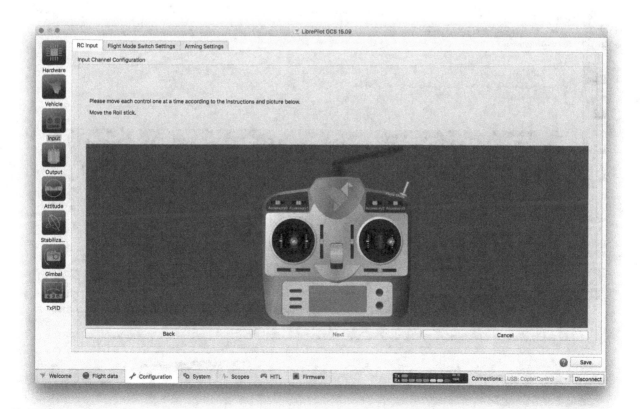

FIGURE 23-5   Roll check.

Next, LibrePilot will have you confirm the pitch, as shown in Figure 23-6.

Finally, LibrePilot will have you confirm the yaw control, as shown in Figure 23-7.

Now that you've set your primary controls, you'll set your "Flight Mode" switch. This switch turns off self-balancing mode. As you learn to fly, leave on self-balance mode. To choose a self-balance switch, press the menu button on the controller. Use the controls to click on "Function Setup" and then "Aux. Channels." Under "Channel 5," choose the switch or knob you want to use. Each physical switch on the transmitter is labeled. Now toggle that switch so LibrePilot can assign it to "Flight Mode" (Figure 23-8).

You can skip through the remaining channels. Finally, LibrePilot will have you move all controls rapidly, which helps LibrePilot learn the maximum outputs of each channel (Figure 23-9).

Finally, LibrePilot will confirm that each channel is working as desired. Make sure none of the channels are inverted and that the movements you make on the transmitter are perfectly mimicked by LibrePilot. Tap the "invert" button to switch any control (Figure 23-10).

You're almost done! Now you just need to choose the arming procedure, as shown in Figure 23-11. On the Cheerson CX-10, you had to move the throttle down, up, and down. On this drone, use the arm command "Yaw Right." "Yaw Right" requires moving the throttle to zero and then pushing it far to the right. Feel free to select any option you prefer.

Both your drone and transmitter are now calibrated, configured, and ready to fly. In the next project, you'll take your drone for its first flight!

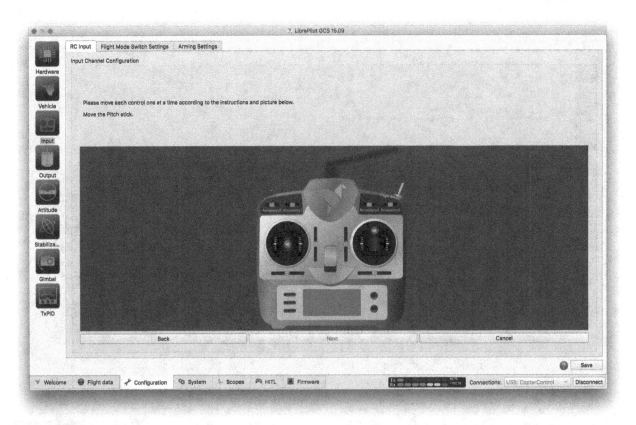

FIGURE 23-6 Pitch check.

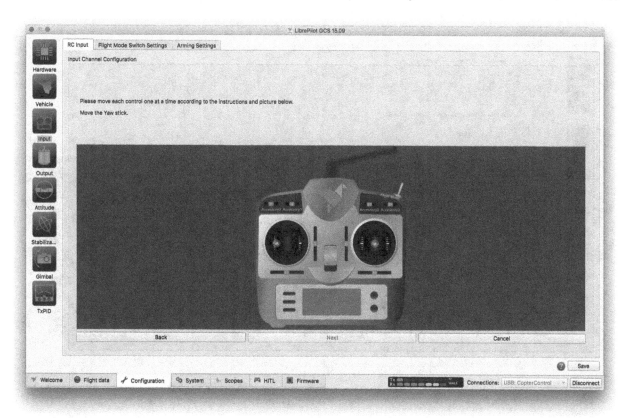

FIGURE 23-7 Yaw check.

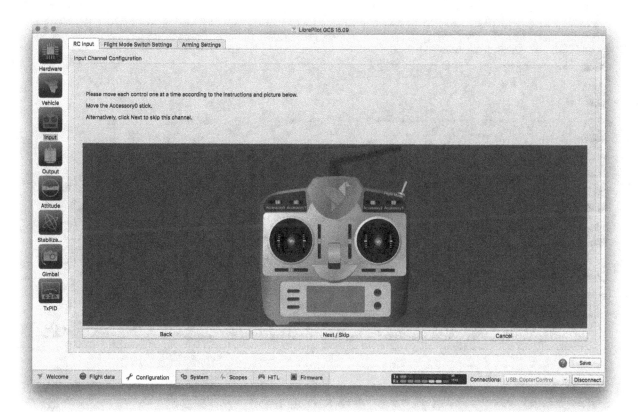

FIGURE 23-8 Flight mode setup.

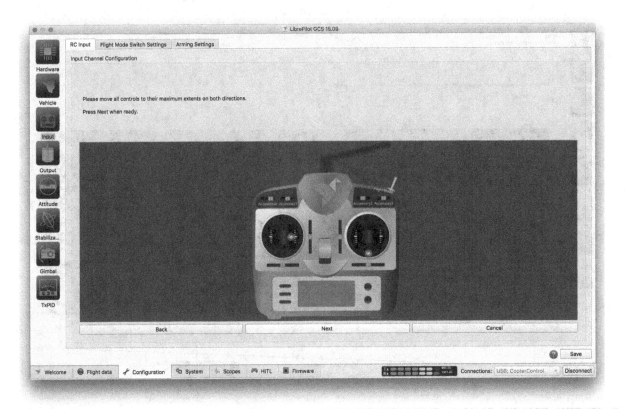

FIGURE 23-9 Maximum movement.

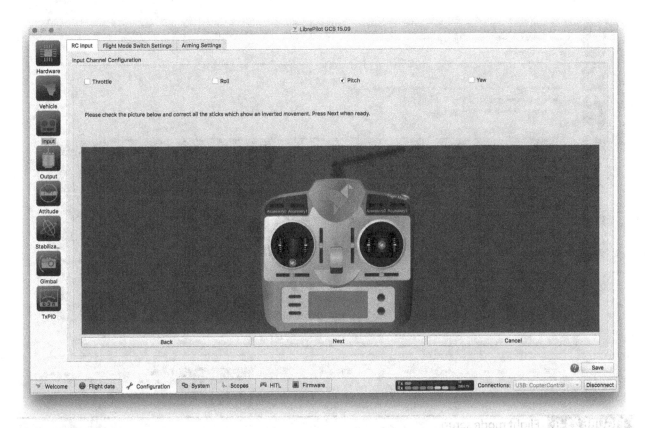

FIGURE 23-10 Invert option.

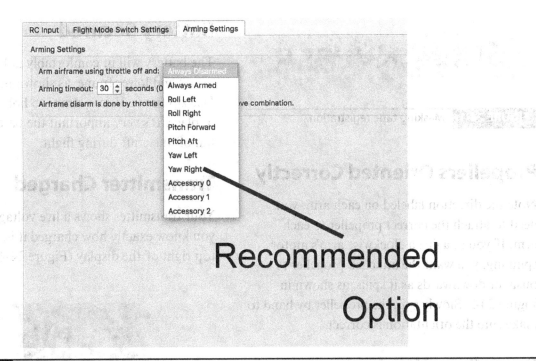

RC Input | Flight Mode Switch Settings | **Arming Settings**

Arming Settings

Arm airframe using throttle off and: | Always Disarmed

Arming timeout: 30 ⬍ seconds (0 | Always Armed

Airframe disarm is done by throttle | Roll Left
                                        | Roll Right
                                        | Pitch Forward
                                        | Pitch Aft
                                        | Yaw Left
                                        | Yaw Right
                                        | Accessory 0
                                        | Accessory 1
                                        | Accessory 2

Recommended
Option

FIGURE 23-11 Arming options.

PROJECT 24

First Flight

Keeping safety in mind, you'll power up the drone and take off for the very first time. After a short flight, you'll practice a controlled landing.

The time has come for your first flight. Before you start flying, make a preflight checklist:

- FAA registration
- Propellers oriented correctly
- Propellers attached
- Battery charged
- Battery secured
- Transmitter charged
- Soft launch and landing zone

FAA Registration

Remember, you must fix your FAA registration serial number to your drone before any outdoor flight. As shown in Figure 24-1, you can attach a piece of masking tape to your drone with your FAA number. If you want a more permanent solution, write the number on a piece of paper and seal it to the aircraft with clear plastic tape or clear nail polish.

SERIAL NUMBER

FIGURE 24-1 Masking tape registration.

Propellers Oriented Correctly

Note the direction labeled on each arm: you need to attach the correct propeller to each arm. If you imagine a clockwise arm's motor spinning, you want the attached propeller to push air downwards as it spins, as shown in Figure 24-2. Slowly spin the propeller by hand to make sure the orientation is correct.

Propellers Attached

The motor shaft nuts unscrew by hand. After placing the propeller on the shaft, tighten the nut. When the motors spin, the screw will self-tighten further.

Battery Charged

It's always important to fly with a full battery. Since you have no way of knowing the battery's current charge during a flight, always err on the side of caution.

Battery Secured

The battery will fit comfortably in between the upper and lower frame, as shown in Figure 24-3. Use zip ties or Velcro straps to hold the battery in place. It's very important the battery does not fall off the craft during flight.

Transmitter Charged

Your transmitter shows a live voltage readout, so you know exactly how charged it is. Look at the top right of the display (Figure 24-4).

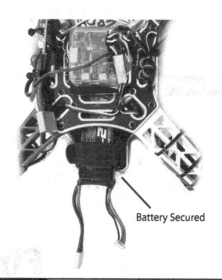

Battery Secured

FIGURE 24-3 Battery secured to quadcopter.

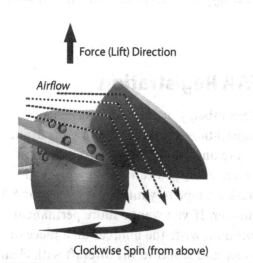

Force (Lift) Direction

Airflow

Clockwise Spin (from above)

FIGURE 24-2 Propeller orientation.

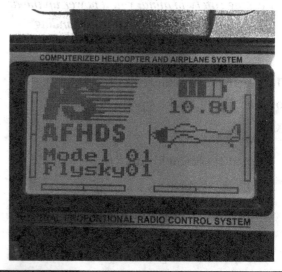

FIGURE 24-4 Transmitter battery capacity.

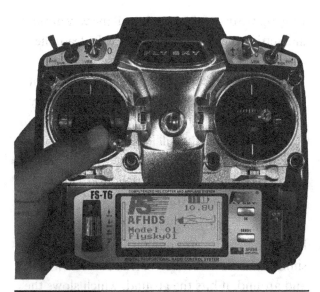

FIGURE 24-5 Arm the quadcopter.

Soft Launch and Landing Zone

Make sure you take off from a soft and flat area. This flight is your first—you don't want to break your quadcopter. Grass is usually best. Dirt and sand are reasonable alternatives, but debris can become lodged in the brushless motor, causing issues.

It's time to get flying! Plug the battery into the quadcopter. Step back five to six feet and turn on the transmitter. Bring the throttle to zero and yaw right, as shown in Figure 24-5.

Keep the pitch and roll controls centered as you slowly increase the throttle. Bring the quadcopter several feet off the ground and very gently adjust the pitch and roll. You'll notice that the controls are sensitive. Be careful not to push too hard—your drone is capable of flying at extremely high speeds and can damage property and people.

As you get a hang for the sensitivity, bring yaw into the picture and try rotating the aircraft. Make sure you remember your orientation; you don't have lights to guide you like you did on the Cheerson CX-10.

Bring your quadcopter in for a soft landing. You'll notice as you get one to two feet off the ground, the quadcopter will drop quickly and then jump back up despite the throttle not changing. This flight pattern is known as "ground effect."

Ground effect creates more lift on an airfoil while it reduces drag, and it comes into play when the airfoil is operated anywhere between the ground and approximately one wingspan above it. In airplanes, ground effect is most often encountered right before landing, when the wings are low to the ground, but the wheels haven't yet touched. From inside an airplane, this feels like floating along the runway just above the ground, even though the throttle is pulled all the way back, and the engine is producing no meaningful power.

Extra lift is produced for the airspeed of an airfoil in ground effect because the air below the wings is squashed against the ground, creating higher pressure below the wings than when operating at the same airspeed out of ground effect. Whenever an airfoil produces lift, it exerts an equal and opposite reaction force on the air below. This reaction force pushes the air underneath the airfoil downwards, causing the air molecules to be pushed more closely together. When air molecules are closer together than the more spread-out air molecules surrounding them, they are at a higher *air pressure* than the others. When operating in midair, the higher air pressure created underneath an airfoil is dissipated by simply pushing some air molecules out of the way to make room for the higher-pressure air molecules to spread out and return to the ambient air pressure. In ground effect, however, those same air molecules cannot be pushed out of the way because the ground is below! The airfoil generates lift, creating a downward reaction force on the air below it. That air immediately below the wing pushes downwards on the air below *it*, and so on, until the air that's "touching" the ground is being pushed downwards from above but can't go anywhere. The ground is receiving a downward force from the airfoil that has been passed through the air from the airfoil to the ground. What happens? Newton's third law happens!

The ground reacts to this force and applies an upwards force on the air that's passed by the air molecules all the way back up to the airfoil. As a result, the air immediately beneath the airfoil increases in pressure rather dissipating into the surrounding air. With a higher pressure beneath the airfoil and the same pressure as always above it, the airfoil produces extra lift. But that's not all.

Wing tip vortices come into play. Wing tip vortices are—simply put—the movement of air around the tip of an airfoil. If you've ever seen iron filings used to visualize magnetic field lines, wing tip vortices are like their air-pressure equivalent. Magnetic field lines are the lines along which the magnetic polarity changes: it moves from positive to negative, gradually, along the field line. The iron filings were magnetized beforehand, so they align themselves along the magnetic field lines. Since the air pressure below the airfoil is higher than the air pressure above the airfoil when it's producing lift, the air pressure must move from higher to lower, just like a magnetic field moves from positive to negative along its field lines. Accordingly, the air between an area of high pressure and an area of low pressure has a *pressure gradient* of gradually changing pressures in between. Below an airfoil (out of ground effect) a pressure gradient extends from the higher-pressure air immediately below the airfoil, downwards to the surrounding air, which is at a lower pressure. At the tip of an airfoil, a pressure gradient extends from the high-pressure air immediately below the airfoil, around the tip to the lower pressure air immediately above it. As a result, air spins in the same direction. This spinning air is a *wing tip vortex*, which occurs at the tips of airplane wings and propellers, helicopter rotors, and quadcopter propellers. The problem with wing tip vortices is the spinning air imposes a downward force on the upper wing surface, reducing the amount of lift generated by the wing. Many commercial airliners use "winglets," which look like fins at the tips of the wings, to stop the

wing tip vortices from rotating all the way around and hitting the upper wing surface. Instead, the force of the vortices is applied to the winglet. The force on the left winglet acts in the direction opposite that of the winglet on the right, so they cancel each other out, and the airplane remains in stable flight. Now, the full length of each wing is able to produce optimal lift, rather than some of it being compromised by the downward force of undisturbed wingtip vortices.

For an airplane in ground effect that doesn't have winglets, the ground performs a very similar function. As the vortex moves down and around, it hits the ground, which slows the air's rotation. This vortex greatly reduces the downward force on the wing tip, maximizing lift while also reducing drag due to the lower air pressure on the upper surface.

As you've noticed, quadcopters experience ground effect like any other aircraft. As they approach the ground, the propellers push air downwards. The ground then reacts by exerting an upward force on the air, which in turn creates higher air pressure underneath the props, allowing them to generate more lift. The tips of the prop blades generate wing tip vortices that are blocked by the ground, maximizing lift and minimizing drag on the blades. As a result, when you descend with a constant throttle setting, your rate of descent will slow when the copter gets low enough to enter ground effect. Extra lift slows the rate of descent, while the reduced drag allows the prop blades to spin more quickly, generating even more lift. As a result, the quadcopter may simply float in ground effect and be impossible to land using the same power setting that was used for the descent. To land, the throttle must be reduced, reducing the RPM of the props, thus reducing lift generated and allowing the quadcopter to settle to the ground.

Congratulations! You've completed your first drone flight of your completely custom-made drone. In the next project, you'll tune your drone's sensitivity for an even smoother flight.

PID Values

Aerospace engineering is back in full force! You'll learn the basics of proportional-integral-derivative (PID) controllers. With your newfound knowledge, you'll optimize the sensitivity and self-balancing abilities of your drone.

Quadcopters are typically stabilized using *PID control*—a type of control system that's optimized for accuracy, precision, and rate of stabilization.

Before you can understand PID control, you need to define accuracy and precision.

Accuracy is the ability to center a process over a target value. When in the process of throwing darts at the bullseye, you might throw darts all around the dartboard. Despite this, if the darts are spread evenly around the bullseye, your throw can be considered "accurate." In control of quadcopters, the accuracy of the control system is the measurement of how well the axis of control sits around a target value. For example, if a stabilization system desires a pitch of 0 degrees when there is no control input, and the quadcopter swings back and forth from +5 degrees to −1 degrees (with an average pitch of +2 degrees), then the stabilization process is less accurate than if the copter swings from +6 degrees to −6 degrees (with an average pitch of 0 degrees).

Precision is the measure of ability to hit the same process value every time. In darts, you might throw all of your darts within an inch of each other at the top-left of the dartboard. This wouldn't be accurate if your target is the bullseye, but the fact that your darts land close together means that your throw can be considered "precise." Similarly, the quadcopter

that swings from +5 degrees to −1 degrees (a range of motion of 6 degrees) is more precise than the copter that swings from +6 degrees to −6 degrees (a range of motion of 12 degrees).

Rate of stabilization is a measure of the time it takes for a system to stabilize to a target value. If control pressure is relieved by a pilot while a drone is pitched forwards at +20 degrees, and the control system is set to stabilize at 0 degrees, then the system will automatically create pitch control pressure to restore the pitch angle to zero degrees. If too much control pressure is added by the system, then the copter will swing past 0 degrees of pitch to −10 degrees. The control system will then create opposite pitch control pressure, pushing it to +6 degrees, etc. The speed at which these swings (*oscillations*) approach (*converge*) the target value is the "rate of stabilization." Generally speaking, the shorter the time, the better the rate.

The performance of a quadcopter's control system—that is its ability to achieve sufficient accuracy, position, and rate of stabilization needed for stable flight—is maximized through proper tuning of the PID system. So what *is* "PID," and what does it mean to tune it?

The PID controller reads from a sensor (such as an accelerometer, which detects the direction of gravity relative to the quadcopter) the true value of a *process variable* at the current moment in time. For this example, we'll use pitch as the process variable but in reality the roll, yaw, rate of ascent/descent, GPS position or altitude, or any other measureable value may be used as the process variable. Assume our true pitch is +7 degrees at this instant. The controller then subtracts the true pitch from the *target* pitch, which may be 0 degrees in a stabilization system. The result, which is −7 degrees in this case, is called the process *error*: the difference between the true value and the target value at the current moment in time. The error is then passed through the three PID filters—the *proportional*, *integral*, and *derivative*—which have a set multiplier, or *gain*, for each one. With the error filtered three separate times and a gain applied to each, the three outputs together produce a *control value* that causes the quadcopter's motion to react to the current error. This reaction can lead to either *convergence* (an increase in stability) or *divergence* (a decrease in stability), depending on the tuning of the PID system. Tuning simply involves adjusting the gain for each of the P, I, and D values.

Proportional (P) is simply the error at the current moment, multiplied by the user-set P gain. A higher P gain will increase the strength of a quadcopter's response to any control error. Doubling the gain will double the strength of the response. If the P gain is set too low, then the quadcopter will have a very low rate of stabilization, if it manages to stabilize at all. If the P gain is set to a negative value, the copter will move further away from the target value, causing it to spin out of control and crash. If the P gain is set too high, then it will far overcompensate for the error causing lots of back and forth oscillations and a slower rate of stabilization. Alternatively, a P gain that is too high may cause the copter to react to the error so strongly that it swings through the target 0 degree value and keeps rotating until it's upside down, at which point it may lose control. This is called an *unstable* system—one that does not converge to a target value but instead diverges away from it. Pretend that the error at the current sampling

time is −7 degrees and the gain is set properly. In this case, the error at the next sampling time might be −3 degrees, then −0.5 degrees, then +0.5 degrees. As you continue forwards in time, a P-only controller with correct gain will converge around the target value but will continue swinging backwards and forwards across it, never settling perfectly to the target value itself. This is an accurate, but imprecise system.

Integral (I) is the compound error over time—that is the sum of some number of errors calculated for each of the preceding time samples. It's a measure of the system's convergence: the larger the integral, the further the system is from converging. For example, a quadcopter that swings back and forth from −4 degrees to +4 degrees will have lower integral values through all phases of the oscillation than a quadcopter swinging from −7 to +7 degrees. The higher the integral value, the greater the strength of the control response. While the P-part of the control system will oscillate about the target value due to an excessive control response on either side of it, the I-part will strip out the excess control response of the P-part and allow the system to fully converge. This type of system is both accurate and precise but requires that the P and I gains are set correctly for the copter. An I-gain too high will cause the system to diverge from the target value, decreasing the copter's stability and causing it to lose control. The P-part may be set high enough to fight excessive I-gain, but the rate of stabilization will drop immensely, and full stability will never be reached. An I-gain too low will slow the rate of stabilization to that of the P-part alone, and possibly prevent the system from ever converging (depending on the ratio of the P-gain to the I-gain).

Derivative (D) is a measure of the difference between the current error and the previous error, taken from the current sampling time and the one immediately preceding it. It's a measure of the direction that the system is moving relative to the target value. If the copter is rotating quickly toward the target pitch of 0 degrees, then the derivative will be large and negative. If the copter is rotating quickly away from 0 degrees of pitch, the derivative will be large and positive. As a result, the D-part of PID control opposes the direction of motion with respect to the target value: that is to say that the control pressure applied to rotate the pitch toward the target value will be reduced by the negative D value, and the specific amount that the control pressure is reduced is proportionate to the speed of rotation toward the target value. The faster it's going, the more the control pressure is reduced, causing the speed of rotation to decelerate and hopefully reduce (or prevent) overshoot. When the D-gain is set correctly, the D-part of PID control limits overshoot to improve the rate of stabilization.

It's just as important to set the gain of each part relative to every other as it is to set the gain of every part relative to zero. Even though the P-, I-, and D-gain settings all differ from one quadcopter to the next, to the same performance, you must remember that there's no concrete answer to the question, "What is the perfect PID setting for my quadcopter?" The "perfect" gain settings for a quadcopter used for one application might be terrible gain settings for that same quadcopter used for a different application. For example, you might use your quadcopter for filming one day, when you'll want it to move slowly and smoothly, so you might do any or all of lowering the P-gain, lowering the I-gain, and increasing the D-gain. All will lead to a smoother-moving system with a slower rate of stabilization, provided that the gains are not adjusted too far! The next day you might want to use that same quadcopter for racing, when you want it to stabilize as quickly as possible, and you care much less about how smoothly it moves. For this change you might increase P-gain, increase I-gain, and slightly decrease D-gain relative to the settings used for filming, again without going too far and turning

your quadcopter into a divergent system that loses control. Control tuning is an *art* in every sense of the word. As with all other forms of art, practice makes perfect!

Easy PID Tuning

The quadcopter you're building uses PID as well. When the drone is in self-stabilization mode and needs to course-correct, the CC3D uses the PID equation. In your initial flight in the previous section, you may have noticed the drone "shake" or oscillate when you stopped controlling the pitch/roll, and the quadcopter tried to stabilize. Adjusting the PID settings will help the quadcopter stabilize more quickly and allow you to have tighter control while flying.

You have many different ways to set the PID values. Since this drone is not a high-performance racing drone, you can use the simplest solution: LibrePilot's built-in EasyTune setting.

To set your PID values with EasyTune, you'll need the following items:

- Quadcopter
- Fully charged battery
- Fully charged transmitter
- USB-Mini Cable
- Laptop with LibrePilot installed

First, plug your drone into your computer with the USB-Mini cable and open LibrePilot. Open the "Configuration" tab and click on "Input" as shown in Figure 25-1.

For this project, you'll want to change the auxiliary inputs on your transmitter to correspond with the two rotating knobs. As shown in Figure 25-2, navigate to Function and Auxiliary Input and set your Channels 5 and 6 to operate via knob control.

On the Input tab click "Start Transmitter Setup Wizard." You've already gone through the setup wizard once before. Do the same thing,

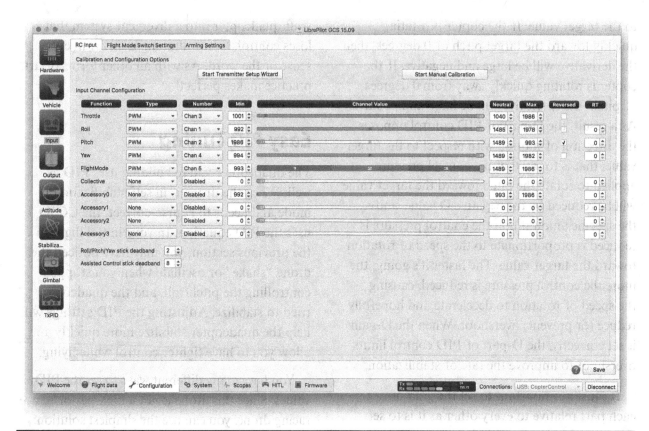

FIGURE 25-1 Input screen.

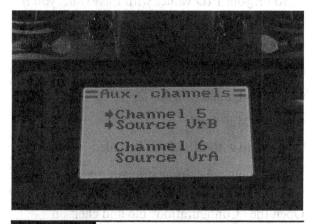

FIGURE 25-2 Transmitter auxiliary settings.

except with some minor changes. Remember to set the arming command back to Yaw Right.

- Flight mode: click "Next/Skip"
- Accessory0: rotate the knob you set to Channel 5
- Accessory1: rotate the knob you set to Channel 6

Hit Save. Unplug, wait 10 seconds, and then re-plug the USB-Mini cable and LiPo battery to restart the CC3D.

Now, navigate to the TxPID tab as shown in Figure 25-3.

The TxPID tab lets you control stabilization values of your quadcopter during flight by adjusting the transmitter's knobs. LibrePilot uses these stabilization values to compute the PID values you need.

Check the "Enable TxPID box." Hit Save. You must now power cycle the flight computer. Unplug the USB cable and LiPo battery, count to 10, and plug both back in (Figure 25-4).

On the TxPID tab, configure the settings shown in Table 25-1.

In addition, change "Update Mode" to "Always." Leave all other settings as their defaults. Click "Save" to save the settings. Go ahead

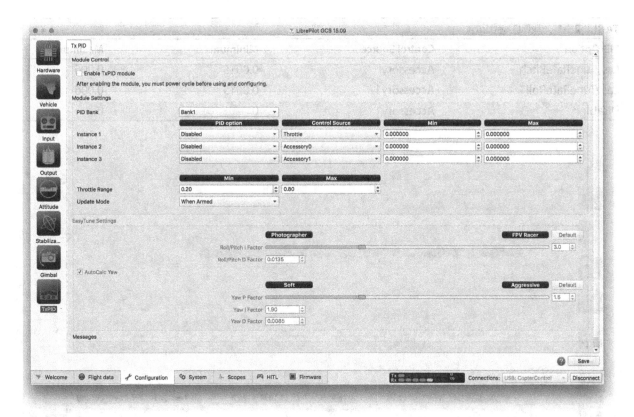

FIGURE 25-3   TxPID screen.

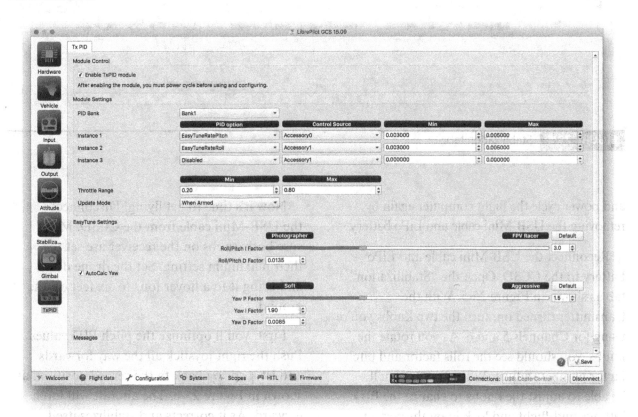

FIGURE 25-4   TxPID screen with some modifications.

TABLE 25-1   TxPID Settings

PID Option	Control Source	Minimum	Maximum
EasyTuneRatePitch	Accessory0	0.0015	0.006
EasyTuneRateRoll	Accessory1	0.0015	0.006
Disabled	Accessory1	0	0

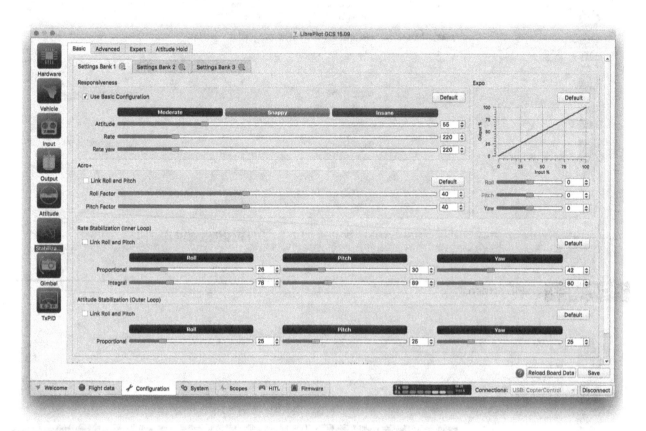

FIGURE 25-5   Stabilization tab.

and power cycle the flight computer again by removing the USB-Mini cable and LiPo battery.

Reconnect the USB-Mini cable and LiPo battery to the CC3D. Open the "Stabilization" tab as shown in Figure 25-5. With the transmitter turned on, turn the two knobs you're using for Channels 5 and 6. As you rotate the knobs, you should see the rolls factor and pitch factor numbers adjust. When you fly, you'll be able to adjust the knobs to change the PID settings mid-flight and lock in on the perfect configuration.

Now it's time to get flying! Disconnect the USB-Mini cable from the CC3D. Make sure both knobs on the receiver are set to their maximum setting. Set the drone down and bring it to a hover four to six feet off the ground.

First, you'll optimize the pitch PID values. Push the right joystick all the way forwards and let it spring back to the zeroed position, as shown in Figure 25-6. The aircraft will lurch forward. As it corrects and stabilizes itself, watch what happens. Does the drone wiggle

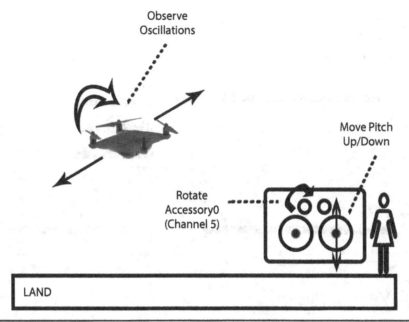

Observe
Oscillations

Move Pitch
Up/Down

Rotate
Accessory0
(Channel 5)

LAND

FIGURE 25-6 Test the pitch.

back and forth? Does it slowly restabilize? Now turn the Channel 5 knob slightly and repeat. Keep turning the knob until the drone quickly restabilizes itself without shaking back and forth.

Second, repeat the above steps but with the Channel 6 knob and the roll (sideways motion). Adjust the knob until the drone course-corrects quickly without over correcting its motion.

Once you're comfortable with the position of both knobs, practice flying your drone, which should feel much more responsive and easier to fly. Bring your drone in for a smooth landing but don't disconnect the battery or adjust the knobs.

Plug your CC3D back into your computer with the LiPo still plugged in. In the Stabilization tab on LibrePilot, you'll see the PID values adjust as soon as the CC3D is reconnected. If you click on "Advanced," you can see the actual PID values as shown in Figure 25-7. Write

these down. These values are unique to your aircraft.

Now you need to save those settings. Turn off TxPID mode, save your settings, and reset the transmitter by going through the Transmitter Setup Wizard once again. Set up the transmitter like a normal transmitter—no need to use the control knobs any more. You may need to change the Aux. Control settings on the transmitter itself as well. Save these changes and reboot the CC3D.

When the CC3D powers back up, go to the Stabilization tab in LibrePilot. Manually enter the PID values you wrote down in the previous step. Save those settings and power cycle the flight computer to double check that your values are indeed saved.

Your PID values are now optimized, you're ready to fly. In the next section, you'll work on adding a variety of accessories from spotlights to GoPros. Fly safely!

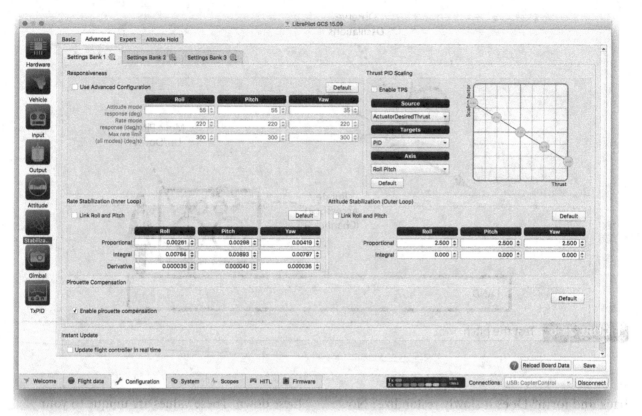

FIGURE 25-7   PID values.

SECTION SIX

Payload Fun

Now you're ready to add some exciting features to your drone: LED lights, GoPro mounts, self-balancing camera gimbals, and more!

SECTION SIX

Payload Fun

High Heels

Your drone deserves some height! You'll create and attach some basic "legs" that will lift your drone's chassis higher off the ground, giving you room to mount a variety of accessories.

The drone you've built is extremely versatile. From your initial flights, you'll notice that the quadcopter is very stable, has a decent amount of power, and is extraordinarily durable. Of course, try to avoid crashing it, but if (when) you do, the drone will usually survive.

Your drone works as a platform for two key reasons: its power and its size. During practice flights, try powering the throttle up to 100 percent while the drone is still stationary. The drone will shoot into the sky at a high rate of speed. If you add more weight to the drone, it will slow down but still be agile enough to fly. The frame's 450 mm size is also perfect for mounting accessories. The frame is big enough to hold action cameras, lights, and other accessories.

To make the most room for accessories, you need to slightly modify the drone's frame. Right now, you don't have much room to mount components under the drone—the clearance between the bottom plate of the frame and the ground is only about half an inch.

Luckily, there's an easy way to add additional height. Quadcopter landing gear is fairly standard and comes in many varieties; the most common types are extended legs, landing skids, and retractable landing gear, as shown in Figure 26-1.

Each type of landing gear has pros and cons, as shown in Table 26-1.

In this chapter, you'll attach landing skids to the drone. Most landing gear is fairly simple to

Landing Skids

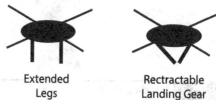

Extended Legs

Rectractable Landing Gear

FIGURE 26-1 Most common types of landing gear.

TABLE 26-1 Landing Gear Pros and Cons

Landing Gear Type	Pros	Cons
Extended legs	Very lightweight	Breaks easily, gets stuck in soft terrain
Landing skids	Very strong	Moderate weight
Retractable landing gear	Folds out of the way during flight and storage	Heavy weight Requires extra transmitter channel

attach, but landing skids are usually the most versatile solution (Figure 26-2).

These landing skids are specifically for these types of frames. You can find them on Amazon at **diydronebook.com**. To attach the landing gear, you need to screw into place the two main support structures, as shown in Figures 26-3 and 26-4.

Once you've tightened the two leg sets, connect the sets with the skid bar. The final setup should look similar to Figure 26-5.

Now that you've attached the landing skids, you have room to put payloads under the main part of your drone. In the next set of projects, you'll strap equipment to the underside of the quadcopter (Figure 26-6).

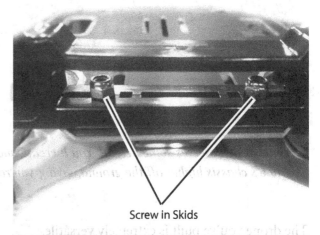

Screw in Skids

FIGURE 26-4   Landing skids screwed to bottom frame.

FIGURE 26-5   Landing skids attached.

FIGURE 26-2   Landing skids.

FIGURE 26-3   Landing skid components.

FIGURE 26-6   Quadcopter ready for flight.

PROJECT 27

Let There be Light!

In this project, you'll attach a series of LED lights and a radio-controlled switch to the drone. You'll be able to fly around in the dark, with the drone illuminating the way!

Now that the landing gear is in place, you have a platform to mount accessories to the quadcopter. To assist with directionality, you can install a spotlight to the front of the quadcopter. This spotlight will shine a bright light in the forward direction, helping you identify which way is forward and letting you fly at night!

For this project, you'll need two components as shown in Table 27-1.

In addition to the LED and switch (Figure 27-1), you'll need your soldering iron and solder (Figure 27-2).

The switch is a very simple circuit with a PWM cable and two wires. When the PWM cable receives a particular input, the two wires allow electricity to flow between them. When the cable doesn't receive a signal, the flow of electricity is blocked. By default, the switch is off. Figure 27-3 shows the schematic of the switch.

The LED is very simple: it has two wires, just like the LEDs from the flashlight project earlier in the book. Remember, polarity odes matter on LEDs: it's important to know which wire is

positive and which is negative. Usually black is negative (ground), and red is positive. The LED you ordered is designed to run off of 12 V, which is the battery's output. Therefore, if you hook up the LED directly to the drone's power distribution board or battery, it will turn on whenever the battery is plugged into a power source.

FIGURE 27-1 Switch.

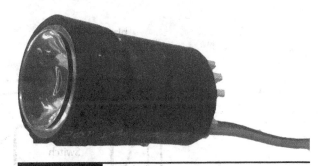

FIGURE 27-2 LED light.

TABLE 27-1 Light Components List	
Part	Details
Receiver-controlled switch	Give the transmitter control over a circuit
LED spotlight	Bright LED light powered by 12 V

REMINDER! Specific part listings are available at **diydronebook.com**.

However, it's way cooler to control the LED via your transmitter. To do that, you'll integrate the LED with the switch as shown in Figure 27-4.

Solder the negative wire from the LED to any negative connection on the lower frame/power distribution board. Next, solder the positive wire from the LED to one of the two wires coming from the switch. Solder the other switch wire to any positive connection on the lower frame/ power distribution board.

Finally, plug the PWM cable directly into Channel 5 or Channel 6 on your receiver. You can unplug the existing PWM cable in that

port—it's unused. Keep in mind the settings on your transmitter and the button triggers that channel.

Now that everything is hooked up, mount the LED onto the quadcopter in whatever direction you see fit. You can have it shine down like a spotlight or forward like a headlamp, as shown in Figure 27-5. The choice is yours!

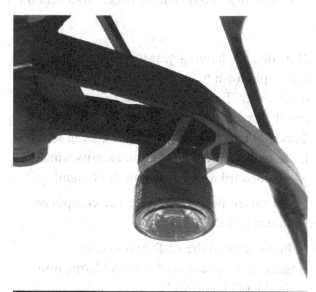

FIGURE 27-5 LED secured to frame.

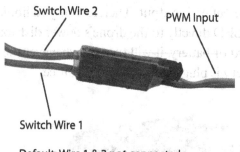

Default: Wire 1 & 2 **not** connected
On State: Wire 1 & 2 connected

FIGURE 27-3 Basics of the switch.

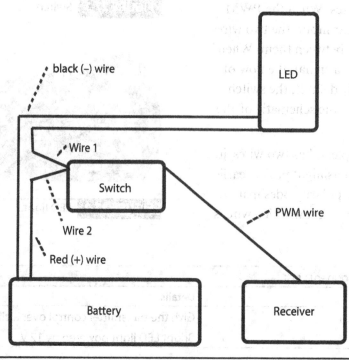

FIGURE 27-4 LED and switch setup.

Power on your quadcopter and transmitter. Flip the appropriate channel switch and watch the LED turn on! As you continue to master your drone, think of other uses for similar switches. You can install more switches, and they can trigger anything, from other lights and electronic gadgets to whatever you can imagine.

In the next project, you'll take your quadcopter to the next level by mounting an action camera to it. You'll be able to film your flights.

PROJECT 28

Filming

You'll mount a standard action camera (GoPro, in this case) directly to the drone to capture first-person, in-flight footage.

How cool would it be to see what your drone sees? Well now you can! Now that your drone has several inches of clearance, you'll easily be able to mount a camera on its underside.

As you've noticed by now, any additional payload or weight on the aircraft not only changes how long it can stay airborne, but also how it flies. You should always optimize for the lightest components possible. In the case of a camera, standard action cameras such as a GoPro tend to be a good fit (Figure 28-1).

Now you need to choose the perfect spot to mount your GoPro. The best way to mount removable accessories is to use standard Velcro straps, as shown in Figure 28-2.

These Velcro straps are strong enough to hold accessories like cameras in place but also flexible enough to be removed when necessary.

When you decide how to mount your camera, keep in mind that, since it will be fixed to the frame, every single movement your drone takes will be reflected in the camera footage. This can be a positive or negative consequence, depending on your intention.

To get a quadcopter point of view angle, mount the GoPro onto the front landing skid facing forwards. Using Velcro, secure the GoPro as shown in Figure 28-3.

Now you'll test it! When you're ready to fly, always turn on the camera first, then turn on the

FIGURE 28-1 GoPro sports camera.

FIGURE 28-2 Velcro straps.

FIGURE 28-3 GoPro secured to drone.

drone, then arm it. Never approach the drone when it's on or armed—remember, safety first! The camera will record a few seconds of extra footage, but you can edit that out later.

Let'er fly! Have a fun outdoor flight and keep in mind the camera's orientation. Try to

fly in a way to get the best possible footage while keeping in mind the orientation of the drone.

After the flight, review the footage. Does it seem wobbly or jittery? Don't worry: that's an easy fix you'll take care of in the next project by attaching what's known as a camera gimbal.

PROJECT 29

Wobbly Woes

Camera footage too wobbly? In this project, you'll learn about two-dimensional and three-dimensional gimbals and how they stabilize camera footage.

Camera gimbals have been around for quite some time. You'll find a lot of variety—from simple shock-absorbing structures and weighted supports to actively controlled, multidimensional gimbals.

In this project, you'll mount a multidimensional gimbal to the quadcopter. In general, gimbals are simple devices: using onboard sensors, they counteract any motion

to keep a camera steady and pointing in the same direction. In practice, they're an advanced piece of technology capable of producing high-quality cinematic footage. To keep the camera steady, gimbals use a combination of gyroscope/accelerometer sensors and brushless motors. The motors counteract the motion caused by the quadcopter and any outside forces like wind.

These gimbals can move in a variety of dimensions. They can tilt the camera up/down, twist it sideways, or rotate the camera. The more dimensions, the steadier the footage, but the more expensive the gimbal, the more it weighs. Most drone accessories will have a tradeoff between functionality and weight.

In this project, you'll use a two-dimensional gimbal (Figure 29-1). The gimbal you'll be using comes integrated with sensors, motors, and camera straps. The gimbal is available on Amazon at **diydronebook.com**.

As shown in Figure 29-2, you can see the two motors and sensor board.

This gimbal keeps things simple: all you need to do is mount it to the quadcopter and provide it some power. Using the pack of extra PWM cables, create a power extension port. Cut the cable in half and solder the positive (red) wire to the positive battery input on the power distribution board (lower frame). Do the same for the negative (black) wire on the negative battery input.

On the gimbal, attach the power cable to the other end of the PWM cable. This setup will allow you to attach/remove the gimbal as needed. The setup should be similar to Figure 29-3.

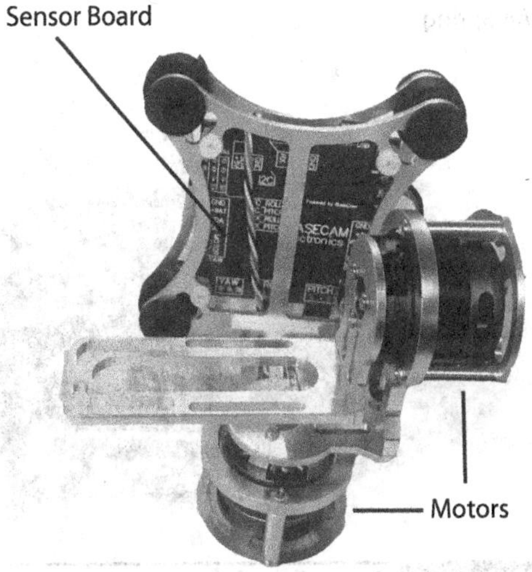

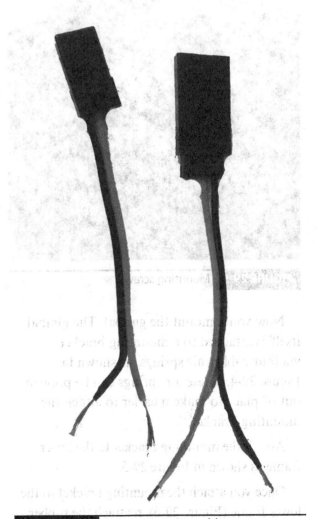

FIGURE 29-2 Motors and sensor board.

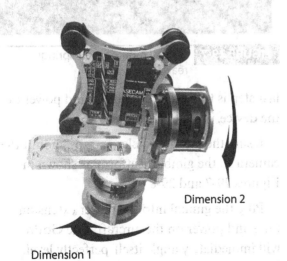

FIGURE 29-1 Two-dimensional camera gimbal.

FIGURE 29-3 Power extension cable setup.

Air Spring

FIGURE 29-4 Air springs.

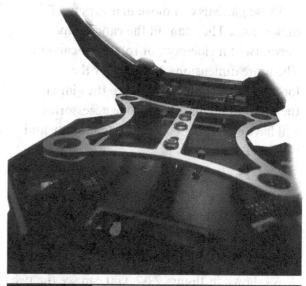

FIGURE 29-6 Mounting bracket.

FIGURE 29-5 Mounting screws.

FIGURE 29-7 Gimbal mounted, air springs reattached.

Now you'll mount the gimbal. The gimbal itself is attached to a mounting bracket via four rubber air springs, as shown in Figure 29-4. These air springs can be popped out of place to make it easier to attach the mounting bracket.

Attach the mounting bracket to the lower frame as shown in Figure 29-5.

Once you attach the mounting bracket to the lower frame (Figure 29-6), reattach the rubber air springs. Your gimbal is now mounted. The last step is to attach the camera and power on the device.

Using the included Velcro straps, mount the camera to the gimbal platform as shown in Figures 29-7 and 29-8.

Plug the gimbal into the power extension plug and power on the aircraft. The GoPro will immediately angle itself perfectly level. Try moving and tilting the quadcopter, and the GoPro will remain completely stationary.

FIGURE 29-8   Gimbal mounted with camera.

Now try flying your drone! Compare the footage you've shot with the gimbal to the footage from directly mounting the camera on the frame. You'll notice a massive difference.

There's one other difference: weight and thrust. You will notice that your drone does not stay airborne as long and feels a bit less agile. Since you added additional weight and an extra power draw (the gimbal sensors and motors), the battery capacity is draining much faster.

In the next project, you'll explore mounting other accessories and swapping out components for other higher-performance parts that will increase the thrust, agility, and flying time of your quadcopter.

PROJECT 30

Ideas Galore

You'll explore how to tune-up the drone to carry more objects and accessories.

Tune Up

From the previous projects, you may have started to feel the limits of your drone's design. The drone wouldn't stay airborne as long, controls would feel sluggish, and the propellers would need to spin more quickly to generate the same lift. Luckily, all of those are easily solvable.

You should remember that quadcopters are extraordinarily modular. Without any soldering, you can swap out the propellers, brushless motors, accessories, flight computer, receiver, and much more.

In this project, you'll make two changes to the quadcopter. You'll upgrade the brushless motors

and increase the battery's capacity. You'll need the parts specified in Table 30-1.

Previously, you were using 935 KV motors. Remember, that means for every one volt of electricity, the motor spins 935 times per minute. The new motors spin 2300 times a minute, per volt! These new brushless motors spin over twice as quickly, which in turn generates significantly more thrust (Figure 30-1). The battery features an additional 600 mAh of capacity, meaning longer flight times.

Upgrading your drone to these new components is easy. Simply unscrew and unplug all four existing brushless motors. Keep in mind their current orientation (clockwise versus

TABLE 30-1 Parts for This Project		
Quantity	**Item**	**Description**
2 CW, 2 CCW (comes in a set)	Brushless motor	2300 KV brushless motors
12	Banana plugs	Male, gold-plated
2 CW, 2 CCW	Propellers	8040 propellers
1	Battery	3300 mAh LiPo battery

REMINDER!   All parts are available on Amazon at **diydronebook.com**.

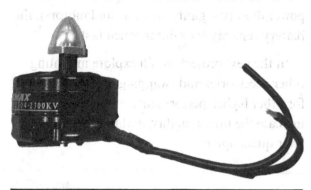

FIGURE 30-1   2300 KV Brushless motors.

counterclockwise). Go ahead and screw the new motors in place. The new brushless motors do not come attached to the male gold-plated banana plugs. Simply solder them in place. There are three plugs per wire. Then attach each banana plug into the ESCs.

Now it's time to attach the propellers. These new propellers have an adjustable shaft diameter. By inserting different spacers, you can get them to work with a variety of motors. Test the right shaft diameter, install the propeller, and tighten the propeller nut.

Go ahead and power up your quadcopter. If any of the motors spin the incorrect way, simply

swap any two of the motor-ESC connections until the motor spins correctly.

Install the larger battery. This change is fairly simple—it just requires plugging in the new, higher-capacity LiPo.

Once your new motors and battery are installed, take your drone for a flight. You may need to retune the drone if it seems to be oscillating or is too slow to respond to your commands.

Now that your drone can carry more weight, here are some ideas for payloads:

- Paper airplane launcher: Connect a servo to one of the receiver channels that releases a paper airplane on command.

- Parachute man dropper: Connect a servo to one of the receiver channels that releases a parachute man on command.

- Package carrier: Put a small lockbox on the underside of the drone and fly items to your friends.

- Sensor kingdom: Attach data loggers such as barometric pressure and wind speed sensors to your drone to measure your flight.

SECTION SEVEN
First-Person Flying

Your ultimate drone add-on is the ability to fly in the first person! You'll explore the world of first-person-view (FPV) drones, including everything from high-speed drone racing to virtual reality goggles.

Equipment Basics

In this project, you'll explore the basic components required for FPV drone flying. You'll examine the pros and cons of different cameras, video transmitters, video receivers, and virtual reality goggles.

So far, you've learned the basics of flight, how to assemble a drone from scratch, and how to fly it. In this section, you'll explore an entirely new perspective on quadcopter flying. So far, your perspective has been fixed to your position on the ground. You've watched the drone from afar, have understood its position in space, and now know how to maneuver it.

But with FPV flying, you sit in the pilot seat. Your perspective is shifted to the drone's perspective, and you fly as if you're onboard the drone itself. To fly in this manner, you need to see what the drone sees. Luckily, we have the technology to make this happen.

A popular subcategory inside the hobbyist drone market is FPV racing drones. FPV means the pilot wears goggles with a video screen so that he or she literally can see what the drone is seeing. These drones are different from your average consumer drone because they're designed to fly much more quickly—around 80 miles per hour! They also lack consumer-friendly features like GPS sensors and aren't built with a focus on design. But what they lack in features and aesthetics they make up for in power. Racing drones are a totally different beast compared to your typical consumer drone, and they're much harder to fly (Figure 31-1).

All racing drones are built differently, but the below description will give you a general idea of how exactly they differ from consumer drones. First, since racing drones often collide with

FIGURE 31-1 Horizon Hobby's vortex racing drone.

obstacles at high speeds and need to be durable enough to withstand the impact, they're typically all built out of heavy-duty material like carbon fiber. Second, most racing drones lack a plastic enclosure or shell, meaning they often look very bare boned. This reduces the overall weight, allows for maximum speed, and provides a stable platform to add or adjust components.

Why would a hobbyist want to add or adjust components on his or her racing drone? The best racing drones are often custom-built by their owners. Similar to NASCAR, high-performance drone racing leagues have gotten too competitive for a hobbyist to simply race with an off the shelf drone. As a result, serious drone racing hobbyists will often buy and modify a high-performance racing drone, or just buy a "blank" racing drone platform and add their own components.

While these components differ based on a pilot's personal preferences, two of the most important parts for FPV drones are the camera and transmitter.

The camera is what captures the video fed into the pilot's FPV goggles, and the transmitter is what sends the signal from the drone to the pilot. But cameras on racing drones are quite different than cameras on a consumer drone. Instead of being high-quality cameras designed to capture great aerial scenery, drone-racing cameras are rather lightweight and low quality, designed only to stream a live video to the pilot. While it's possible to record this video for post-race playback, most pilots will choose to attach a higher-quality camera like a GoPro if they want to capture footage to be watched after the race. The transmitter is essentially an antenna sticking out of the drone. Instead of relying on Wi-Fi to transmit the live video like consumer drones, racing drones rely on older analog video transmission via radio signals. While this technology has been around for a while, it arguably streams more quickly and is more reliable than Wi-Fi.

On the ground, each pilot needs a receiver to pick up the video transmission from the drone and a pair of FPV "goggles," so they can see what the drone is seeing. These goggles cover a pilot's entire field of view, which means they can't see anything around them. While the goggles might look strange, they ensure the pilot

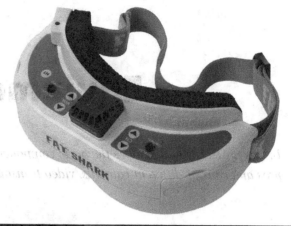

FIGURE 31-2   FPV goggles from Fat Shark.

can focus all of his or her attention to the race (Figure 31-2).

Now that you know how a racing drone works, you should know where and why people race them. Drone racing, which started as a small-time weekend hobby, has quickly evolved into an international sport (Figure 31-3).

Professional racing leagues all over the world attract thousands of fans and spectators. Corporations that are hoping to attract attention in the growing industry are even sponsoring these races. Races can take place both indoors and outdoors, and drones usually race on a track (Figure 31-4).

An amateur race may use cones and tall poles to map out a track, while high-stakes races will incorporate high-energy LED lights that make the course look like a scene from TRON, as shown in Figure 31-5.

FIGURE 31-3   Drone race course from XDC 2 @ Zappos. Photo by Jeff Gale of GreenGale Publishing.

FIGURE 31-4   Another drone race course from XDC 2 @ Zappos. Photo by Jeff Gale of GreenGale Publishing.

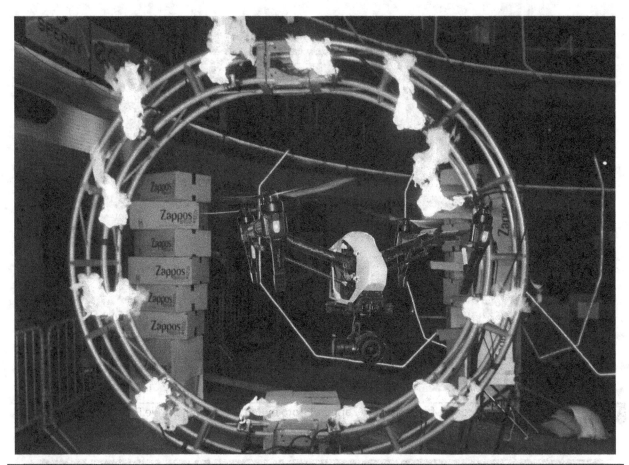

FIGURE 31-5   DJI Inspire flying through fire ring XDC 2 @ Zappos. Photo by Jeff Gale of GreenGale Publishing.

PROJECT 32

Broadcasting Live!

You'll now choose proper components and mount them onto your existing drone. You'll power everything up and make sure the live stream is working as expected.

Why would you want to strap on goggles and see what your drone is seeing or even live stream at all? Well, first off, if you want to competitively race your drone, utilizing FPV technology is a necessity. And while this project will mainly cover setting up live streaming for first-person racing, it's not the only reason someone may want to get live video footage from a drone.

Imagine trying to take a picture of someone with your eyes closed. It would probably be difficult, and the picture would turn out poorly! Well now imagine taking a photo with your eyes closed when you are hundreds of feet away from the camera, which is what it's like to take photos or videos on a drone that doesn't provide the operator with a live view of what the

TABLE 32-1   Parts for This Project

Quantity	Item	Description
1	Camera	*Comes in bundle*
1	Transmitter with antenna	*Comes in bundle*
1	Receiver with antenna	*Comes in bundle*
1	Monitor	*Comes in bundle*
1	Headset	*Comes in bundle*
1	Battery	LiPo battery with XT60 plug (any mAh)

REMINDER!   All parts are available on Amazon at **diydronebook.com**.

camera is seeing. Live-streamed video, whether it's streamed via Wi-Fi or a radio, is an integral part of any amateur or professional drone photographer's toolset. Without a live stream there's no way to see how the footage looks until the drone is back on the ground, at which point altering the shot would require an entirely new flight. At the end of the day and no matter what your drone's purpose, live video streaming is an indispensable feature that all modern drone operators should utilize.

In this project, you'll start with a simple live stream. You'll choose the FPV components, mount them to the drone, and test out the system. You'll need the parts shown in Table 32-1.

Most of the FPV components are beginner components and come in a bundle package. In a later project you'll look at optional upgrades. Each component in the bundle can be upgraded individually.

When the bundle arrives, identify the camera and transmitter as shown in Figures 32-1 and 32-2.

A variety of cables come in the bundle. Find the set that fits the camera and transmitter plug. An additional red wire will connect to your quadcopter's power distribution board.

As shown in Figure 32-3, be sure to attach the antenna labeled "Tx" (stands

FIGURE 32-1   FPV camera.

for transmitter) to the antenna plug on the transmitter board.

Solder the red power plug to the power distribution board. You can also create a removable solution by using the extra PWM cables from the previous section. Either way, you need to power your FPV setup.

So how does this setup actually work? When the transmitter is powered on, it receives signal from the attached camera. It sends that signal out over the specified channel on the transmitter via the antenna. Any receiver with the right channel and frequency can pick up the video signal.

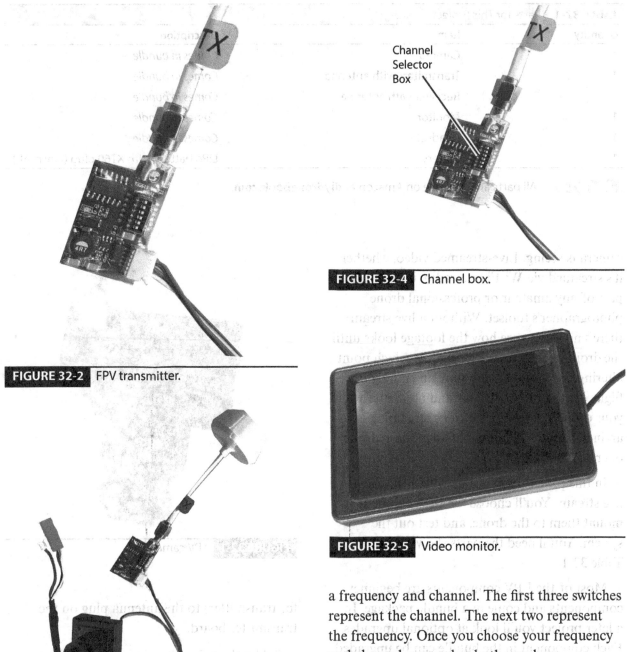

FIGURE 32-2   FPV transmitter.

FIGURE 32-3   Quadcopter FPV components.

FIGURE 32-4   Channel box.

Channel Selector Box

FIGURE 32-5   Video monitor.

Picking a Channel

Take a close look at your receiver. You'll notice a box of very tiny switches. Using the manual that came with the receiver, flip the switches to select a frequency and channel. The first three switches represent the channel. The next two represent the frequency. Once you choose your frequency and channel, remember both numbers (Figure 32-4).

Next, you need to set up the receiver. The receiver is powered by a LiPo battery. You can use a spare quadcopter battery to power it. Get out the video monitor, receiver, and extra wires, as shown in Figures 32-5 and 32-6.

Make sure the receiver has the "Rx" (receiver) antenna attached. In addition, you'll need the cable with the XT60 plug and the cable with the red/yellow/white composite tips.

FIGURE 32-7 Cable setup.

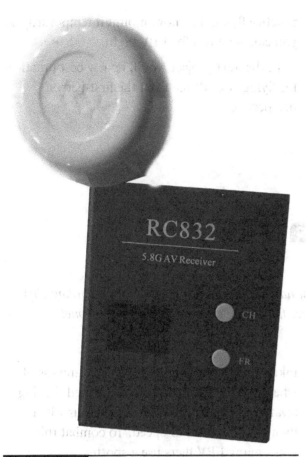

FIGURE 32-6 Receiver with Rx antenna.

As shown in Figure 32-7, connect the power cable with the XT60 plug to the receiver and into the monitor's red composite plug. This cable feeds power to both the receiver and the monitor. Next, take the composite adapter (the cable with the white and yellow composite plugs) and plug the yellow end from the monitor into the cable. Plug the other end of the cable into the receiver. Leave the white plug from the monitor disconnected. The white plug is for audio, and plugging it in will simply cause interference.

Now you'll test the system. Power on both your quadcopter and FPV receiver setup. Remember the frequency and channel numbers from early—go ahead and cycle through the options on the receiver until you reach the correct channel. Once you do, you should see the monitor change from "no signal" to showing the camera's perspective.

FIGURE 32-8 Completed headset.

Finally, assemble the virtual reality headset. Instructions are included and will require some glue. When complete, the headset should look similar to Figure 32-8.

Finally, secure the components to the drone. Mount them with zip ties to avoid the propellers and make sure the camera is facing the front of the drone. Don't worry about the mounting angle too much—you'll need to adjust it as you practice flying. For now, mount it temporarily so you can get a feel for FPV flight.

In the next project, you'll review best practices for flying your drone from the first-person perspective.

PROJECT 33

FPV Safety

Before you take your craft airborne, you should know additional safety considerations regarding FPV flight, including the role of a spotter and best practices for having the most fun with the drone!

FPV flight is very different from normal flight and eye-of-sight perspective. Instead, it's easy to fly your FPV craft outside of your visual boundary. Since you can keep flying as long as you have a video signal, you must be extra careful to obey the FAA's quadcopter rules and always remain within a visual perimeter of the quadcopter, as shown in Figure 33-1.

In addition, when flying an FPV drone, you can easily lose sight of where the takeoff zone is located. If you lose spatial awareness of the takeoff zone, you might also lose awareness of where you, as the operator, are located. Losing awareness can be extremely dangerous when flying at high rates of speed. To combat this issue, many FPV fliers use a spotter.

Spotters maintain a visual line of sight with the aircraft and tell the pilot information about the drone's current position. The spotter may

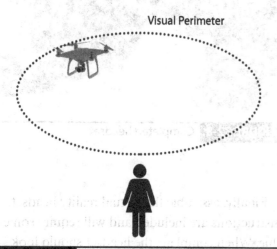

FIGURE 33-1 Maintaining the visual perimeter.

Visual Perimeter

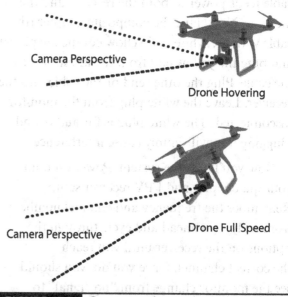

Camera Perspective

Drone Hovering

Camera Perspective

Drone Full Speed

FIGURE 33-2 Camera angle, unadjusted.

call out commands such as "pull up" if the craft is too low, or "turn back" if the quadcopter is exiting the visual perimeter.

Finally, the camera angle is very important. If you're planning on flying pure-FPV, you want to make sure you're always able to see in front of you. Now try flying your FPV-equipped drone. You'll notice as you accelerate forward, the drone will pitch and tilt the camera in a new direction (Figure 33-2).

To combat this effect, most FPV pilots keep the quadcopter always in motion. With the quadcopter always moving forward, the drone is permanently tilted forward, allowing the operator to choose a standard camera angle. Adjust the angle, tilting the camera toward the top of the quadcopter. Try flying full speed ahead forward and determine if the angle is acceptable during full speed flight. If so, use additional Zip Ties to further fix the camera to the quadcopter. If not, adjust the camera and try again.

In the next project, you'll actually practice flying the quadcopter using the FPV perspective.

PROJECT 34

Take Control

In this project, you'll fly your FPV drone and work on some basic maneuvers.

Your FPV system should now be set up and ready for flight. Before you take your drone out to the field, test your system at home and make sure the video quality is acceptable. The transmission occurs over the 5.8 GHz channel, which is frequently used in household devices. Outdoors, the signal will be vastly improved. The antennas are also best operated when they have a line of sight connection to one another. That means flying the drone around a building and out of view of the receiver will cause a loss of video signal. Keep these facts in mind during your first flight!

Follow the below checklist for any FPV flight:

1. Quadcopter battery fully charged
2. FPV receiver battery fully charged
3. FPV visual transmission working
4. FPV camera mounted at correct angle
5. Understanding of visual perimeter at flight area

For this first flight, you should bring a spotter. If you don't have a spotter, carefully observe your flight area. Look for landmarks such as trees or buildings to get an understanding of the field's visual perimeter. Once you're familiar with the area, situate yourself in a position next to an obvious landmark. You always want to make sure you can find your own location when operating your FPV quadcopter.

Set your drone a few feet away from you and power on all systems. Plug in your FPV headset and receiver and make sure the video transmission is working. Since the drone is stationary, the camera will be pointed at an

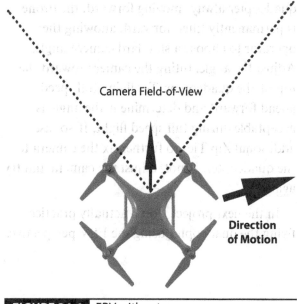

FIGURE 34-1   FPV without yaw.

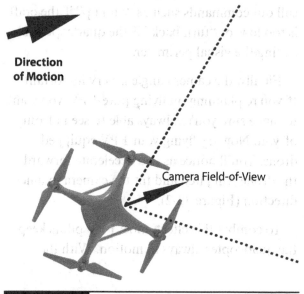

FIGURE 34-2   FPV with yaw.

awkwardly high angle. Don't worry—as soon as you start moving, your quadcopter will pitch forward, fixing the angle.

Perform one final check of the area and put on your FPV headset. Give the drone some altitude and pitch forward. As soon as you start moving forward, the camera angle should give you a good perspective on your multicopter's heading. As you fly, be sure to incorporate yaw (aircraft rotation) (Figures 34-1 and 34-2).

Start by completing laps around your flying area. Make sure you can always spot where you're sitting—the last thing you want to do is fly the drone into yourself! As you become more comfortable with the quadcopter, keep in mind the drone's altitude. It's easy to lose track of how high (or low) you are in FPV mode. It will take a considerable amount of practice to get a better feeling of altitude.

When you begin to feel even more ambitious, try flying through some obstacles! Common obstacles include:

- Figure-eight pattern around football goal posts
- Fly under the foliage of trees
- Fly in circles around lightposts
- Spot yourself flying the drone and land next to your takeoff zone

FPV flying can be a very frustrating experience at first. Your drone is very durable, and almost all of the parts are replaceable; it's okay to experience an occasional crash. Remember, practice makes perfect. As you continue to fly, you'll become more and more comfortable with your quadcopter.

In the next project, you'll explore variations of FPV flying and system upgrades you can make.

FPV 2.0

FPV is addicting, isn't it? You'll explore upgrade options, higher-quality video signals, and ways to record your in-flight experience.

Once you get the hang of FPV flying, it becomes extremely addicting. Flying from the first-person perspective lets you control your drone in a whole new way.

Just like the rest of your quadcopter, most of the parts are interchangeable. First, you should review core FPV system components:

- Transmitter
- Camera
- Receiver
- Heads-up display

The transmitter uses the 5.8 GHz band, which is standard for FPV systems. You can, however, upgrade the antenna. A selection of upgraded antennas is available on **diydronebook.com**. These antennas screw into the same coaxial cable port and can broadcast with increased range or directionality.

The camera you're currently using is great for high-speed FPV flying: it might not be the highest quality, but it's durable and lightweight. If you're more interested in cinematographic flying, you can stream live from your GoPro or other action camera. Simply replace the transmitter power and camera cable with the GoPro cable. The cable plugs into the side of a GoPro and transmits the GoPro's view live to your FPV receiver, as shown in Figure 35-1.

The receiver and headset in the bundle are fine for FPV flying but may become burdensome

FIGURE 35-1 FPV via GoPro.

after a while. The cables are not that long, and the system is heavy, requiring you to hunch over and lean down when flying in FPV mode. To combat this issue, many companies have created high-quality, all-in-one, headset-and-receiver devices. These systems are more expensive but have additional features such as flight recording, fan-cooled monitors, and integrated telemetry systems. A selection of these headsets is available at **diydronebook.com**.

If you really want to take your FPV flying to the next level, you can even check out rigs that allow you to fly in virtual reality (VR).

Rising in popularity over the last few years, VR is the newest trend in consumer electronics. Companies like Oculus (owned by Facebook) and HTC have released VR headsets that give users a 360-degree immersive experience (Figure 35-2). These headsets are outfitted with advanced head tracking and motion sensing technology, so you literally feel like you're actually experiencing whatever content you're watching.

While the VR space is still young, a few collaborations between the VR and drone industry have already occurred. One expected-use case that's already being tested is technology that allows drone operators to replace FPV goggles with VR headsets. Since VR gives users a 360-degree view, piloting a FPV drone in VR would be much more realistic than using a 2D goggle display. VR could also be used to create applications for drone pilots to practice flying in a safe environment. A user could strap on a VR headset, grab a drone controller, and feel like

FIGURE 35-2   Oculus Rift virtual reality headset.

they're actually outside practicing their piloting skills. As VR technology advances, expect to see a lot more innovation happening in the drone and VR space.

SECTION EIGHT

Onwards and Upwards

Now that you've mastered the basics of aerospace engineering and drone flight, you should look forward to finding future avenues to explore.

Intelligence

In this project, you'll look at commercial solutions for "intelligent flight."

You might think the flight computer you've been using in your drone is advanced, but just wait until you hear how artificial intelligence, or AI, is helping drones basically fly themselves. Over the past few years, companies have been working on building drones that don't require human interaction. The potential benefits of an autonomous drone are tremendous. Not only can autonomous drones operate with greater accuracy and efficiency, but they can also operate themselves, meaning one human could potentially oversee dozens or even hundreds of automated drones all at once.

One of the most promising areas of drone AI is photography. Imagine throwing up a pocket-sized drone before heading down your favorite ski slope. As you descend the mountain, the drone follows you, automatically capturing shots of all your best moves. The drone can even automatically avoid obstacles, so it doesn't accidently run into a tree or mountain. Some AI drones even let the user instruct it to perform specific functions then carry out the instructions without any human interaction. One such drone is Solo, which is made by 3D Robotics. While Solo can be flown manually, it also allows the pilot to specify in advance what type of shot they want taken, and the drone's computer will take over from there to capture the perfect shot. Even cooler is that Solo lets pilots split the work with the flight computer. As a result, you could choose to have the AI either act as a pilot or cameraman. For example, if you need a very specific camera shot, you could tell Solo to fly itself on a predetermined path so you can concentrate on operating the camera.

Although Solo's AI features could allow it to be totally autonomous, it's still technically supposed to be flown with a controller. But other photography drones are currently being developed that won't even require a controller because they'll always control themselves. For example, the Lily camera drone can literally just be thrown in the air, and it will navigate itself from there. The drone will automatically know where you are and follow and record whatever is happening. As advanced as these drones seem, they're really just in the infancy of both the drone industry and artificial intelligence research. One day in the not-too-distant future, a world might exist where both people and cargo are transported by autonomous drones in the sky.

> We are entering a new age of less intrusive and more personal computing driven by A.I. and embedded systems. We are a few years away from technology so intuitive and invisible that it would feel like magic today. Lily camera is one of the first devices that reflects this new technology paradigm—ultra-personal computing.
>
> –Antoine Balaresque, CEO of Lily

In addition to photography drones, delivery drones that will also use elements of artificial intelligence to help them fly without human control are being developed. As mentioned earlier, companies like Amazon are developing drones that will eventually be able to delivery packages directly to your doorstep in fewer than 30 minutes. Since Amazon hopes to eventually have thousands of drones in the sky all making deliveries at once, the company needs to develop a way for the drones to fly autonomously without hiring thousands of pilots.

Artificial intelligence and automation are also changing how the military flies drones. While the secret nature of the US military's research and development activities prohibit civilians from knowing details about the military's projects, multiple branches of the US military have long been using drones for combat and reconnaissance. What is known is that soldiers still control most drones, but they're rapidly becoming more and more autonomous as the necessary technology is invented. Aircraft-style drones have already landed on an aircraft carrier without any human control.

It's pretty clear that artificial intelligence and autonomous flight are the future of the drone industry. But is this a good thing? Yes, because as you've learned, removing the human element from drone flight will allow many more drones to fly at once much more efficiently. If every commercial drone in the air is autonomous, they will always be connected via wireless networks and be able to communicate with each other regarding location, direction, and speed. This will dramatically reduce the risks of in-air collisions, and make the skies much safer. So does this mean the days of hobbyist drone flights are coming to a close? Absolutely not! Hobbyists will always be able to fly drones manually. But in the not too distant future, airspace could be divided between autonomous drones and human-operated drones. Amazon has proposed a solution where airspace below 200 feet is reserved for hobbyist drones without automated sense-and-avoid technology, and the space between 200 and 400 feet is reserved for autonomous drones that can automatically communicate and avoid each other. Ultimately, while there's no way of knowing what exactly the future of autonomous drones holds, the era of both hobbyist and commercial drones is just beginning.

PROJECT 37

Hexacopter? Octocopter? Bigger? Smaller?

Quadcopters are fun. You know what's even more fun? Hexacopters! Octocopters! The more arms the better. You'll explore the pros and cons of different drone frames.

Drones come in a variety of shapes and sizes. In general, when drones need to carry more payload, they need to generate additional lift, and they can accomplish that in a number of ways. However, the most standard method is to increase the number of propellers. A quadcopter becomes a hexacopter or even an octocopter. Of course, more propellers and motors mean more power use. The drone will be heavier, require bigger batteries, and suffer negative effects from more turbulent air.

On the opposite side of the spectrum, drones can be a lot smaller. The Cheerson CX-10 you practiced with earlier is a great example. You can even go a step further and build/customize those miniature drones.

One great example is the Inductrix drone. Similar to the Cheerson, the Inductrix is a quadcopter available from Amazon. Unlike the Cheerson, however, the Inductrix uses more standardized parts that can be upgraded. By making a few modifications, the Inductrix can be made powerful enough to lift an FPV camera (Figure 37-1).

To modify the Inductrix drone, you simply need to replace the four brushed motors and increase the battery size. All of these parts are readily available online, with links to the specific components at **diydronebook.com**.

Once the motors have been swapped out (no soldering required), try test-flying the drone. It will be very responsive and strong enough to carry a payload. You can buy a fully integrated FPV camera and transmitter that's compatible with your existing receiver and headset. The

FPV system can be glued directly to the Inductrix and power cables soldered to the battery leads on the Inductrix's board, as shown in Figures 37-2 and 37-3. Again, instructions are available at **diydronebook.com**.

Once completed, your quadcopter should be ready for indoor FPV flight!

When viewed through your FPV headset, the world looks new. Since the drone is so

FIGURE 37-2   Drone with FPV.

FIGURE 37-3   Drone with FPV.

FIGURE 37-1   Inductrix drone.

FIGURE 37-4   FPV perspective.

small, your house or apartment can now become the perfect FPV drone-racing course. Weave around chairs, under furniture, and more (Figure 37-4).

Want to build the Inductrix FPV drone yourself? Parts and instructions are available at **diydronebook.com**.

This tiny FPV drone is just one example, but it should make you realize the potential of quadcopters. Whether you want to scale up or down, the possibilities truly are endless.

PROJECT 38

Brain Dump

Building drones is great, but there's nothing wrong with buying a ready-to-fly one.

While this book has mainly focused on teaching you how to build your own drone, many drone companies sell preassembled, ready to fly drones. These drones can range from $20 to $5000, and all have different abilities and features. Although tens if not hundreds of different companies make these drones, five drone manufacturers are major industry players today.

The first company is DJI. Founded in China in 2006, DJI makes commercial and recreational drones that specialize in aerial cinematography. DJI has three main families of drones that they produce. The first is the Phantom series (Figure 38-1), which will set you back about $1400 for the newest version. The latest Phantom has a built-in 4K camera with 3-axis motorized gimbal, built-in GPS and vision positioning system so it can avoid obstacles and hover in place, and a swappable 28-minute battery. Essentially, the Phantom is a hobbyist

drone that's powerful enough to also be used for professional cinematography. DJI's second line of drones is the Inspire series, which costs anywhere from $2000 to $3000 depending on which accessories you get. While still technically considered a ready-to-fly consumer drone, the Inspire is powerful enough for almost any professional cinematographer. The drone can

FIGURE 38-1   DJI Phantom.

even support two controllers linked at once, so one pilot can control the drone while the other controls the camera and gimbal. The last family of drone that DJI makes is their industrial flying platforms. These drones have up to eight propellers and are designed to carry a professional DSLR camera. These rigs can cost upwards of $5000 and are definitely only needed by professional aerial photographers.

The next company is 3D Robotics, which was founded in California in 2009. Much smaller than DJI, 3D Robotics is known for producing Solo, a consumer drone that retails for about $1000 (Figure 38-2). While the Solo doesn't have a built-in camera (it was designed specifically for a GoPro), it's advertised as one of the smartest drones ever made. The drone can essentially control itself and comes preprogramed with specific maneuvers and camera moves that it can autonomously perform to capture amazing shots. The drone can also be a virtual two-person film crew—it can automatically fly itself while you control the camera, or automatically control the camera while you concentrate on flying. It can even do both at the same time, so you don't have to control the drone at all!

The third company is Parrot. Parrot was founded as an electronics company in 1994 but recently has devoted a great deal of its attention to the drone industry. While Parrot produced its first drone in 2010, its current flagship product is the Bebop. The Bebop sells for around $350 and

has a built-in 1080p HD camera but no gimbal. The Bebop is definitely more consumer-oriented than DJI or 3D Robotics' drones and can be controlled with just your iPhone. The drone is also very lightweight, weighing only three pounds. That being said, it's a great drone for the price and can capture some great aerial footage (Figure 38-3).

The fourth company is Cheerson, founded in China in 2011. While Cheerson definitely isn't as well known as the previous three companies, it makes a fantastic micro drone called the CX-10, as shown in Figure 38-4. You flew this drone earlier in the book! The tiny drone weighs less than an ounce and is only a few inches large! But the CX-10 has created an entirely new

FIGURE 38-3   Parrot's Bebop drone.

FIGURE 38-2   3D Robotics solo drone.

FIGURE 38-4   CX10 drone.

category of drone, one that can be flown in your own living room. It even has a tiny, low-quality camera built in so you can get some aerial shots inside your house.

The last company worth mentioning is GoPro. While GoPro hasn't yet officially unveiled its drone, the company has announced that it's working on one that should be released by the end of 2016. While details are scarce, it's safe to say that the drone will give the above competitors a run for their money, mainly because of GoPro's expertise in the camera industry. The vast majority of existing drones already use a mounted GoPro camera to capture footage, so it was only a matter of time before the company entered the drone market itself.

Show Off

In this project, you'll learn best practices for editing and sharing quadcopter videos and photos.

So you've built your drone and installed a great camera. Now what? Well, if you're like us, you probably want to take advantage of your new camera in the sky and share some great footage with your friends and family. In this project you'll learn the best way to capture video and share it with the world. Assuming you already have a drone with a working camera, be sure you have the right memory card to save footage while in flight. Nothing is worse than getting ready to hit "record" while you're in the air and then realizing your camera doesn't have a memory card. Most drone cameras accept Micro SD cards, but you should double check before you buy one. In addition, try to buy a card with a lot of storage, at least 8 or 16 GB. While you may end up paying a few extra dollars now, the money will pay off when you can fly confidently knowing you'll never run out of storage space while capturing that perfect shot.

Now that you can save your footage, you should feel confident about what you're doing in the air. While you may be tempted to fly fast when you're recording, a slow flight speed is actually key to capturing great footage for a few reasons. First, drone cameras must frequently adjust their focus and contrast because the sun will often appear in the frame, especially if you're flying in the morning or early afternoon. Flying and turning your drone slowly will allow it ample time to adjust to the different levels of sunlight while in flight. Second, fast speeds often equal shaky drone footage, which definitely isn't professional. If you want your footage to appear professional, as if shot from a cinematic rig, fly your drone slowly and smoothly. If for some reason the footage is too boring when you fly slowly, you can always fast-forward in postproduction to avoid the shaky footage.

Now that you've recorded some awesome footage, you should polish it and share it with the world. The first step is to transfer the footage you recorded from your memory card to a computer. Depending on the type of memory card you have, you'll probably need a USB card reader. Once the card is plugged into your computer, copy the files to your desktop or preferred storage location on the hard drive. Now, you have a few options. You could directly upload this footage to the Internet, but you'll probably want to do some light editing first to polish it. Depending on whether you have a Mac or PC, you can find free video-editing tools available for you to use. Regardless of which tool you choose, it should be able to perform all the functions below.

First, look through your footage and delete the parts you don't want. Since a good drone operator always errs on the side of recording too much footage rather than not enough, there will probably be a decent chunk of footage you can cut out. After you cut out the boring stuff, you can decide if you want to add slow motion or fast-forward effects to the video. Both effects can look great on drone footage, so you should try them both to see what suits you. You also may want to add some music to your flight. The sound from most drone recordings is normally just the propellers spinning, and that gets boring after a while. Add some music to match the theme of your footage to make it more fun to watch. If it's a majestic landscape try adding opera or classical music. If it's a cityscape, modern music may be a better fit. But whatever you choose, you can't go wrong, so have some fun with it!

After you finish editing, you'll want to export the footage so it can be shared. Depending on your software, you'll have multiple export options, but be sure to always export in the highest quality: 4K or 1080p depending on the camera you used. While a high-quality export will take a lot longer (and render a much larger file) than a standard export, it's 100 percent worth the wait. Plus, since you'll be uploading it to a video-sharing platform, you don't have to worry about storing the large file size.

Now that your export is done, let's upload it to somewhere everyone can see. Many video hosting sites exist, but you'll be using YouTube, since a large community of hobbyist drone flights already exist on the site. If you don't have a YouTube account, you should make one now. Even if you already have one, consider making a second one to use exclusively for uploading your drone flights. That way, you'll have a dedicated place on the web only for your drone videos. Uploading to YouTube is rather straightforward, and you can find instructions on the website. While you're on the site, consider searching for some other great drone footage for inspiration. Not only will the footage give you ideas for your next flight, it will also show you what types of drones other hobbyists are flying and building.

PROJECT 40

Airborne!

Now that you're a master drone pilot, you're ready for some final points.

You've learned a lot. You're a registered small-unmanned aerial vehicle pilot, and you know how to fly drones indoors and outside. You can select drone components, assemble one from scratch, and even fly from the quadcopter's first-person perspective.

You'll find many areas within the drone industry to explore: FPV racing, camera platforms, commercial applications, and more. As discussed throughout this book, quadcopters use a fairly standard set of components. Once you build your first drone, you can use many of the components to try out different types of quadcopters. For example, for less than $20, you can buy a 250 mm frame on Amazon and remount the brushless motors, flight computer, and receiver to build a high-end FPV racing drone.

FPV Racing

FPV racing is a growing industry. Find a local group on **Meetup.com** or another community gathering site to meet fellow racers.

Most FPV drones use a 150 or 250 mm frame, both of which are available on Amazon. You will also want to upgrade your camera and headset as mentioned in the previous FPV section.

Camera Platforms

Drones allow photographers to capture completely new perspectives. GoPro and other action cameras offer an inexpensive way to capture high-quality footage. Using a camera gimbal lets you capture smoother footage. You can also explore building larger drones such as a 550 or 650 mm hexacopter or octocopter that will carry heavier payloads like DSLR cameras.

Commercial Applications

Commercial use of drones is strictly governed by local, state, and federal statutes. Always be sure you're operating within the law.

Some common commercial applications of quadcopters include inspecting hard-to-reach areas for insurance or safety purposes. Drones are also increasingly being used for transportation of commercial goods.

Let your mind go wild and try to push the boundaries of the industry. Just remember: always keep safety in the forefront of your mind and obey all local, state, and federal multicopter regulations. Most importantly, always have fun and keep learning!

Index